MARKETING FOREST PRODUCTS

Also published by Miller Freeman for the forest products industry

Computer Control Systems for Log Processing and Lumber Manufacturing
by Ed M. Williston

Dry Land Log Handling and Sorting
by Charles M. Hampston

Farming the Small Forest
by Laurence C. Walker

High Temperature Drying: Enhancing Kiln Operations
by Larry Culpepper

The Logging Business Management Handbook
by Ronald R. Macklin

Logging Practices, revised edition
by Steven Conway

Lumber Manufacturing: The Design and Operation of Sawmills and Planer Mills, revised edition
by Ed M. Williston

Managing Mill Maintenance: The Emerging Realities
by Richard F. Baldwin

Modern Particleboard and Dry-process Fiberboard Manufacturing
by Thomas M. Maloney

Operations Management in the Forest Products Industry
by Richard F. Baldwin

Plywood Manufacturing Practices, second edition
by Richard F. Baldwin

Quality Control in Lumber Manufacturing
edited by Terence D. Brown

Saws: Design, Selection, Operation, Maintenance, second edition
by Ed M. Williston

Small Log Sawmills: Profitable Product Selection, Process Design and Operation
by Ed M. Williston

Timber Cutting Practices, third edition
by Steve Conway

Value-Added Wood Products: Manufacturing and Marketing Strategies
by Ed M. Williston

Directory of the Forest Products Industry

FOREST INDUSTRIES Mill Map

FOREST INDUSTRIES North American Factbook

MARKETING FOREST PRODUCTS

. . . gaining the competitive edge

Dr. Jean Mater
with
M. Scott Mater
Catherine M. Mater

Miller Freeman, Inc.
San Francisco, Atlanta, Boston
Chicago, New York, Brussels

Library of Congress Catalog Card Number: 91-61846
International Standard Book Number: 0-87930-193-7

Printed in the United States of America
91 92 93 94 95 5 4 3 2 1

Contents

Chapter VI To Market, To Market—
Moving the Product From the Factory to the Customer

Chapter VII Promotion—
It Doesn't Have to Cost an Arm and a Leg

Chapter VIII Pricing—
If The Price Is Right

This book is dedicated to

MILTON H. MATER

who foresaw the changes in
the forest products industry.

PREFACE

Comparison of end-of-the-century America with the America of the post World War II era reveals that today we are living in a vastly different world; the entire forest products industry faces a new way of doing business. No longer can we depend on the same customers and the same methods that have sustained our industry for so many years. Many forest products companies have realized that the brightest opportunities for stabilization of the industry hinge on developing new products for specific customer groups.

This book applies new concepts in marketing to the forest products industry. It is addressed to both the newcomer and the experienced marketer. It especially meets the needs of smaller and medium sized companies where marketing is but one of the jobs of the manager who wears many hats.

The entire forest industry has been faced with staggering changes:

The population has changed. The baby boom of the early post World War II years has swept over the country like a tidal wave, bringing with it consumers with values and behavior patterns that differ from those of their parents and grandparents.

In its wake, the tidal wave has left smaller families and an assortment of less traditional households, unimaginable affluence as well as incongruous poverty, and hundreds of new markets. America has become more diverse racially, ethnically, socially, and geographically. The mass market of the 1950s characterized by the popular television show, Ozzie and Harriet, has given way to market fragments, segments, and even smaller "particles."

As the post-war baby boom matures, we are moving from the accent on youth to the unprecedented era of the "graying of America." The population of the rest of the world is also aging, affecting buying behavior in almost every country.

Markets have changed. New home construction has lost its domination as the driving force of much of the forest products industry. Today, with the nation's housing stock getting older, a new force has emerged: Home remodeling, repair, and improvement have surged ahead in creating markets for forest products. The distribution revolution led by giant home-center stores and retail building material chains has created consumers of wood products, rather than just commodity lumber.

Geographical location of the markets has changed. After decades of steady increase, the population of the United States grew only ten percent in the past ten years, and more than half that growth was concentrated in California, Texas, and Florida. The population of the West now exceeds that of the Northeast. Buying power has shifted from the rural areas to the suburbs and metropolitan areas.

Timber resources have decreased. In many parts of the world governments are intervening in timber harvesting. The environmental movement is pressuring wood products manufacturers and consumers not to use a host of suddenly designated "endangered wood species." Indonesia has embargoed the export of unprocessed timber. Rattan, for example, can only be exported as part of manufactured furniture. Indonesia, like many other countries, has switched from being a resource supplier to a product supplier. Competition for forest resources is increasing as the availability of those resources diminishes.

International competition has changed. Both global competition and global opportunities have increased. United States trade associations are introducing American practices in the use of wood to other countries. Meanwhile, other countries are introducing their manufacturing technologies to the United States. Building materials exports have moved from a happy trade surplus to a larger deficit. Almost every American industry has lost global market share in the past decade.

All these changes demand switching from the practice of selling or trading commodity lumber to marketing wood products. The switch is not an easy one. The forest industry is comfortable with processing and selling commodities. Marketing wood products demands much more from a manufacturer. Modern marketing is the creation of products and services to meet human needs and wants. This requires that we identify these needs and wants and understand how they differ from each other in the hundreds of new market segments that have splintered-off from the traditional markets.

New and old products compete for the consumer's attention and buying power. More products than ever crowd the shelves of stores. Consumers and industrial users have more choices and are bombarded with more promotion as producers compete for customers. But with slower population growth, new customers are getting harder to find. Learning how to differentiate, position, promote, distribute, and price the products we make to meet buyers' needs has become as important as adopting new manufacturing technology.

Manufacturers of industrial wood products are faced with the necessity of understanding not only their immediate customers, but their customers' customers.

Marketing pays off in expansion of products, sales, and profits. It offers any innovative, creative business of any size undreamed of opportunities to tap into new and emerging domestic and global markets. But marketing is not without risk; all business ventures are risky. This book is designed to help minimize those risks

To accommodate the fluidity of consumer lifestyles, working habits, and public attitudes, marketing itself has evolved at once into a new science and into a new art. Although marketing is an idea as old as politics, as enduring as courting, and basic as business, anyone who studied marketing ten years—even five years ago—might scarcely recognize it today. New words and phrases and new meanings for old words have crept into the marketing language: megamarketing, packages-of-benefits, lifestyles, niche markets, segmentation, do-it-yourselfers, negative product differentials, home center retailers, the green movement, reverse distribution, house of quality, and high-impact marketing are among them.

Pharmaceuticals, textiles, steel, food, and other industries have also faced foreign competition and fickle consumers. The marketing initiatives that helped them survive suggest possibilities for the forest industry. Not all of their experiences are directly applicable to the wood products industry, but they are included as examples in this book to stir the reader's imagination and generate new ideas.

This is a "how-to," self-help book, accompanied by extensive background for those who want more information. Ideas are summarized in checklists. Throughout the book are details on finding and promoting niche products, lists such as 25 ways to differentiate industrial wood products, comparisons of different types of distribution, and methods of pricing products. Suggestions on simple projects—as how to turn serving coffee and doughnuts into a publicity coup—are added to the lists of ideas for low cost promotion. The concept of *singularization* is introduced as one of the low-cost methods for making wood products stand out from the crowd.

Four appendices are included: there is a glossary explaining many of the terms used in this book, and some of the terms appear in italics for emphasis; a directory of forest industry organizations, which includes the names and addresses of the majority of specialized associations that serve the industry; a list of forest industry trade journals; and a list of marketing reference books and periodicals useful for marketers.

There's no doubt about it. We can't continue to do business in today's world in the same old way. We hope that this book will help the forest industry use new ways to better compete in the 90s.

We express our appreciation to the innovators, risk takers, and investors who exemplify the opportunities of the forest products industry and make it one of the most interesting and valuable businesses in the world. We also thank all those whose knowledge contributed to this book.

Jean Mater
M. Scott Mater and Catherine M. Mater
April 1991

I

MARKETING FOREST PRODUCTS

. . . not like marketing toothpaste

CHAPTER SUMMARY *Marketing is an idea as old as politics, as enduring as courting, and as basic as business. In each, we make a promise to give people what they want and need. Success is measured by winning the vote, gaining acceptance, or making the sale. The marketing premise is that people buy not only products or services; they buy expectations and solutions to problems. Forest industries, as other industries, have traditionally focused on selling the products they make rather than making the products people need, thereby missing opportunities to satisfy expectations. The marketing approach identifies and attempts to meet customer's wants and needs. Marketing offers the wood industry the best opportunity to successfully compete as domestic economic conditions change and opportunities open in international trade.*

Marketing strategies on options for expanding sales form a framework for everyday marketing decisions. Some marketing strategies work best for smaller companies with slim resources; others give larger companies the advantage of economies of scale.

The concept of megamarketing, which combines political power and public relations with marketing techniques, is introduced as a method for gaining entry into otherwise closed markets and for gaining public support for the forest industry.

Note: Many terms that appear in this book are defined in the glossary, which starts on page 263. Some appear in italics for emphasis.

Marketing offers the forest industry the best opportunity to compete in the emerging global economy and to level out the economic peaks and valleys typical of our free enterprise system. No two ways about it: almost all industries are facing the same intensification of competition. Other industries have gone through the strain of finding ways to win a solid place in the competitive sweepstakes. Most have found the answer in marketing. This book gleans from the successes and failures of other industries and innovators in the forest products industry. From these experiences, sawmills, plywood plants, furniture and pallet manufacturers, hardwood dimension producers, manufactured home businesses, and the hundreds of other wood-based businesses may also find the path to becoming successful competitors.

Marketing forest products is like playing Poker. Although luck has a dominant role, experience in the game increases the chance of winning. All players are dealt the same number of cards; the large number of possible winning combinations and the shrewdness of the players in filling a hand makes poker an enticing game. As competition sharpens, players become more innovative. So it is with marketing: Competition is increasing; players are gaining more experience; the game becomes more interesting.

WHAT IS MARKETING?

Marketing has become the buzzword of business, though people appear to have different ideas of what marketing is.

The word marketing is tossed about freely and used loosely. Marketing and markets are often used interchangeably, though they are different. Marketing is the development and distribution of products and services that create and satisfy customers. Markets are the customers. The word Marketer (or the plural, Marketers) refers to persons who are engaged in marketing activities.

Many business managers have the idea that marketing means selling; others think that it means advertising; still others believe it means packaging, and distribution. Ask one manager about a marketing program and the reply is: "I'm heavily into marketing. Why, I've just started a sales training program." Ask another and you'll hear: "We have a terrific promotion program." Still another claims to be marketing because his company now shrink-wraps its products.

All these activities are included in marketing: promotion, a sales training program, a new package are all part of creating and satisfying customers.

Marketing focuses on people, especially those people who have the potential to become customers. Marketing recognizes that people don't buy a product or service; they buy expectations or solutions to their problems. Filling these expectations is what marketing is all about. Modern technology has made possible the production of an increasing variety and number of products and services. Marketers match these available products or services to those customers' expectations.

Marketing includes the act of selling, but selling accomplishes only part of the marketing objective. Figure I-1 compares selling and marketing.

FIGURE I-1

The Difference Between Selling and Marketing

In selling, a company sells what it can make; it focuses on the product. A company that markets makes what it can sell; it focuses on the customer.

A selling-oriented company concentrates on production; a marketing-oriented company concentrates on satisfying customers. A sawmill selling lumber focuses on the product, with little direct relation to specific customer needs.

In selling, prices tend to respond to market demand. In marketing, the producer has a greater control of the price of the product.

Glance at any issue of *Random Lengths* or the *Hardwood Reporter* and note how commodity lumber prices rise and fall with demand. Except for increasing production efficiency and market penetration, producers of commodity lumber and panels have limited bottom-line control. The dilemma familiar to primary manufacturers in the forest industries is that often log and chip prices rise while lumber prices fall.

In selling, packaging is designed to protect the product; in marketing, the product is packaged for the customer's convenience.

Customer service is sometimes considered a necessary evil in selling; it is part of the product in marketing.

Selling frequently uses little strategic planning or feedback from customers. Marketing coordinates planning of all activities—from resource acquisition through production—involved with making and keeping customers.

A company that sells, concentrates on the manufacturing aspects of products. A company that markets, focuses on the design of products to meet a customer's needs and wants. "Engineering in reverse"—beginning with the customer's preferences and designing the product to meet these preferences—is typical of companies with a marketing mentality.

Lumber is usually sold. Products, like cabinets, are marketed. "Value added" products bridge the marketing chasm between commodities and products. Marketing can provide the alchemy that turns commodities into higher value products by meeting specific customer needs.

Note in Figure I-2 how a Georgia-Pacific advertisement in Home Improvement Center (September 1990 issue) meets the needs of its home center customers. The left hand page of the advertisement carries the message with this quote from the manager of a home center store: "I could make even more on moulding if I could buy smaller quantities more often. Is that so hard?" On the right hand side, GP promises: "Order a lot or a little . . . When your moulding arrives, packaged in palletized bundles, it'll be clean and undamaged . . . Easy to handle . . . GP's appealing point-of-sale

FIGURE 1-2
Georgia Pacific Tells How It Meets the Needs of Customers

Source: Home Improvement Center 1990

materials create a display . . . consistent profiles . . . the kind of moulding that keeps customers coming back . . . the simplest way to keep your profits growing."

Price in relation to quantities purchased is not the factor GP uses to position its products against competitive moulding.

Granted, GP is one of the forest industry's giants. Smart smaller wood products companies piggy-back on the marketing know-how of large companies with departments that specialize in marketing. Smaller companies can offer the same product benefits: manageable quantities, palletized bundles, point-of-sale materials, consistent profiles.

IT'S NOT LIKE MARKETING TOOTHPASTE

Marketing wood products is not like marketing toothpaste. Toothpastes are a convenience item, a product that the consumer purchases frequently, with minimum effort. Toothpastes are similar in purpose, contain equivalent ingredients, and compete in a win-lose market. If one toothpaste manufacturer gains market share, another will probably lose some. Differences that separate one toothpaste from another depend on ingenuity to devise features that attract customers: For instance, Crest®

toothpaste became the nation's leading toothpaste by winning an American Dental Association endorsement.

Toothpastes are no longer marketed the way they used to be, either. Marketing toothpaste has heretofore relied heavily on promotion to create brand preference to make the toothpaste more desirable, and positioning to encourage buying it. Brand marketing was previously geared to selling leading brands to millions of shoppers with similar tastes and loyalty to particular brands. Brand marketing created a comfortable world for marketing managers, but today's consumers cannot so easily be sold by brand alone. Toothpaste producers are now dividing their customers into discrete market segments and creating toothpastes calculated to appeal to the different tastes of each segment.

Except for wood toothpicks, matches and similar items, consumer wood products are *shopping items* usually bought on the basis of shopping and comparisons between products. From relatively inexpensive novelty bread boards purchased as souvenirs to a major family investment in furniture, consumers compare how the physical attributes and styling fit their needs, tastes, and budgets before making a purchase.

Family finances are more important in buying products that are considered to be essentially the same. Most children's shoes were once considered to be relatively similar, or as marketers say, homogeneous. No longer. Brand name shoes for running, athletics and other special purposes have moved shoes into the heterogeneous, or essentially different class, aimed at particular market segments and retailing at considerably higher prices.

Product Differentiation, another marketing concept, is the key to converting a product that is similar to other products into a more valuable, different product. When a wood product becomes an identified product with unique characteristics that buyers prize, the product differentiation strategy has succeeded. The buyer makes an effort to find that particular item and is willing to pay a higher price for it. The product meets his wants—another marketing term for "desires," as contrasted with "needs," which are essential. The concept of singularization uses the unique characteristics of each wood species to help ordinary wood products become prized specialty products.

THE *FOUR P'S* OF MARKETING—OLD CONCEPT, NEW ROLES

Marketing influences every activity of a company. Marketing considerations shape the design and functions of the products a company produces, how these products are distributed to customers, the price the company charges, and how customers learn about the products. The business community calls these basics the *four P's* which are: product, place, promotion, and price. The *four P's* are interdependent. Alter one P and the other three are affected. How companies use the *four P's* to increase sales and gain the competitive edge is the essence of this book. Figure I-3 provides an overview of the *four P's* for reference.

The *four P's,* as described in Figure I-3, point the way to new opportunities for the wood products industry.

Products As Packages Of Benefits

Consider the definition of product as a "package of benefits." Studies have shown that today's busy consumers need and want products that are convenient to use. Wood products manufacturers of products for the growing repair and remodeling business are learning to design

FIGURE I-3
The *Four P's* of Marketing—Building Blocks for Marketing Strategy

PRODUCT—In marketing, a product is viewed as a "package of benefits" that includes the physically identified item, accessories, packaging, identification, and services that are essential to using the product. In marketing, warranties, service, instructions, installation, availability of compatible or supplemental products are all part of the product.

PLACE—The product must be physically available to customers when and where they want it—in the right place at the right time. Promotion efforts, a reasonable price, and a desirable product are of little help if the customer does not have ready access to the product at the time he wants to buy it.

PROMOTION—Ralph Waldo Emerson's conclusion that: "If a man can write a better book, preach a better sermon, or make a better mousetrap than his neighbor, . . . the world will make a beaten path to his door" does not reflect real world experience. The world will not find the door or want to make a beaten path without promotion. Promotion informs the market about the product and how it meets the market's needs. Personal selling, sales promotion, advertising, publicity, and point of sale are all used in promotion to arouse interest in the product and create customer preference for the product.

PRICE—Pricing is one of the most creative aspects of marketing. In modern marketing, price reflects the value the customer places on the product. Another buyer motivation that plays a role in the price value is the buyer's desire for status. If the product provides pride or increased status, the buyer may willingly pay more.

products with built-in convenience for application and use. The recent resurgence of hardwood flooring is attributed to the development of new finishes that make hardwood floors as easy to maintain as other types of floor coverings. The increase in the number of women who work outside the home left women with little time for waxing and polishing floors. Adding the benefit of easier maintenance helped handsome hardwood floors to regain favor with homemakers.

The package-of-benefits concept is prompting manufacturers of wood products to join with producers of new finishes and applicators to provide more convenience in applying and using wood products. The new wood preservative that offers protection against termites, ultra-violet rays, mildew and water—all in one coat—enhances the package of benefits of kits for building wood decks. The no-drip squeeze bottle that saves time and mess in applying the rapid-drying new pastel colors to wood cabinets can be sold with cabinets. Each new product that saves steps and is quick and easy to apply and clean up increases the consumer's tendency to purchase wood products, rather than products manufactured from competitive materials.

Distribution As Diversification

Consider the concept of place. *Channels* of distribution, as the marketers call them, are opening new opportunities. Customers may now select furniture from many types of retail establishments instead of only the traditional retail furniture stores and conventional department stores. To purchase an item of furniture, a consumer may now shop in general merchandise outlets, specialty furniture stores, furniture chain stores, mass merchandisers and discount department stores, warehouse showrooms, manufacturer-owned stores, wholesale furniture distributors, mail order catalogs, discount furniture stores, buying clubs, and even through television programs.

The trend of home center stores and retail lumber yards toward the merchandising of projects, rather than selling sticks of lumber, generates opportunities for wood products manufacturers.

Rickel Home Centers, an Eastern chain owned by Supermarkets General Corp., Carteret, New Jersey, decided to exclude lumber from its "HOM" stores. Although most home improvement retailers hold that lumber is an essential stocking unit for a home center store, Rickel tried a store that sells hardware, decor, and seasonal products—but no building materials and lumber. Rickel's decision was based on two strategies: (1) not to compete with the lumber departments of The Home Depot and Home Quarters warehouses and (2) to focus on catering to women shoppers, rather than men. To reach these women, Rickel markets ready-to-use and easy-to-use wood products in place of lumber. Playing up home decorating and repair projects instead of "construction," Rickel stocks wood products that a woman can use as easily as a paint brush. This marketing approach has earned HOM stores the title: "The 'Porsche' of home improvement stores."

Promotion As Opportunity

Consider the new opportunities for promotion of products created by the communications revolution. No company can afford to believe that its product is so necessary and so appealing that consumers will flock to the door. New technology has made it possible to bombard consumers with information about products, making promotion more than ever a key to expanding sales. Even great innovations such as the automobile and the airplane, which created new markets, need promotion. The rise of product quality to the top of the line of product attributes desired by consumers is creating new opportunities for manufacturers. Consumers now demand that the products they buy perform as promised. Products that are "lemons" can't be sweetened by slick advertising. Nor can a company count on surviving by "selling the sizzle instead of the steak," if the steak isn't good in the first place.

Price As Value

Consider the new role of price. Pricing may include mark-ups, discounts, and other inducements to buy, but according to marketing experts, the price the buyer will pay for a product is usually determined by the need or perceived value of the product. This concept is so new that some businesses still price their products according to production costs and cash flow requirements, and other firms set prices on what the market will bear, without considering the product's value to the consumer.

Paul B. Brown writes in *Inc.* magazine (May 1990): "It's fair to say price is probably the least effective arrow in your marketing quiver." Least effective although it may be, price can be an effective marketing tool when used in conjunction with the other *P's*.

DRAWING A MARKETING ROADMAP

The *four P's* are the basic units of marketing roadmaps or strategies. A strategy is a plan that results when a company asks itself: "What kind of company do we want to be? What direction do we want our company to move in the future?"

A marketing roadmap helps keep the company eye on its objectives. It exerts a discipline to assure obtaining greatest value from dollars spent on marketing. A strategy keeps resources focused on accomplishing goals. It provides a sense of direction.

Strategy guides company decisions about which products to make, the manufacturing method, selection of production equipment, the funds to commit to developing a product, the method of distribution, personnel selection, and priorities.

For a demonstration of the value of strategy, consider the success of the New Zealand wood pulp producer, Winstone Pulp International Ltd. (WPI), in penetrating the global pulp and paper industry. Swimming against the tide of brutal competition, WPI devised a strategy that succeeded in marketing its wood pulp products to a rigidly targeted market.

WPI's strategy was based on built-in flexibility to make different grades of wood pulp for a number of different end users, as contrasted with the limited flexibility of other pulp producers. The strategy segmented the world market geographically and targeted sales to the rapidly growing Pacific Rim. WPI's pulp could be used in newsprint, printing and writing paper, paper board, tissue, or fluff. The company concentrated on paper board and tissue grades to best utilize its radiata pine raw material, and used product differentiation to distinguish its pulp from other, largely undifferentiated pulp commodities.

The most important element of WPI's strategy was strong service and technical support to customers. Tailoring production to customers' requirements, WPI adopted the policy of producing only "to order" to assist customers in meeting their end-user requirements.

WPI clearly knew what kind of company it wanted to be and where it was going. The strategy helped WPI succeed in selling all of its production to the Pacific Rim and over 70 percent of its output for paperboard and tissue production, the two areas the company targeted. Businesses are often put off by strategy or planning. Suspicious of any hocus-pocus forecasting, they avoid what they believe may be a time-wasting exercise. They still practice shoot-from-the-hip, defensive marketing, reacting to changes without a plan of where they want to go.

Nevertheless, managers make strategy decisions every day when deliberating such questions as:

- ❑ Shall we invest in a new planer to upgrade our product?
- ❑ Should we bid on this government contract to supply pallets?
- ❑ Should we buy the Midvale factory and diversify our furniture manufacturing capability?

- ❑ Shall we limit production to basic commodity lumber?
- ❑ What wood products should we manufacture? What won't we consider?
- ❑ Shall we aim at the global market?
- ❑ Should we expand out of our current geographical market?
- ❑ Should we market this product to industry?
- ❑ How about trying to enter the consumer market?
- ❑ Do we have the capabilities to manufacture secondary products?
- ❑ Do we have access to adequate timber, or lumber, or chips?
- ❑ Should we expand into the do-it-yourself market?

Companies constantly search for strategies that will improve their competitive advantage, for ways to create and maintain a position in their industry. Strategy decisions may be made at a formal board meeting, at an informal lunch, or even over a friendly cup of coffee or a stop at the water cooler.

Marketing Options

However or wherever the strategy roadmaps are drawn, basic decisions involve deciding where the company is headed, its *mission*, as marketers say, in the business world, and its goals and objectives in accomplishing this mission. A company mission is the backdrop against which it can test decisions. Marketing strategy decisions will probably narrow to one of these four options about products and markets:

1. ***Continue making existing products for existing markets*** Expand sales to increase penetration of these markets by:

- ❑ Attracting new customers
- ❑ Convincing present customers to increase orders
- ❑ Improving existing products
- ❑ Intensifying promotion by enlarging the sales force, increasing advertising, and improving brand recognition
- ❑ Improving quality control
- ❑ Adjusting prices

In carrying out this existing products/existing markets strategy, a company faces further decisions on the competitive position it wants to achieve in the industry and the most effective strategies to use. Figure I-

4 shows some of the further considerations in competition and strategy methods.

FIGURE I-4
Some Decision Choices on Competitive Positions and Strategy Methods

What competitive position does the company want to achieve in the industry?

Become a *market leader,* with the largest share of the market?

Become a *market challenger* and confront the market leader at his own game?

Become a *market follower,* satisfied with a smaller share of the market, and avoid going head-on against stronger competitors.

Skip the competitive position and find a market niche.

STRATEGY METHODS

In pursuing the option of continuing existing products, a company might adopt a *differentiation strategy,* adding features and services that offer the purchaser unique advantages. Products may be differentiated by offering various types of services (guaranties, instructions, assembly tools), packaging, and materials. The firm with creative personnel or specialty engineering skills often finds product differentiation its most effective strategy.

Or a company may select a strategy of cost leadership in which it attempts to become the low cost producer of the product, (without ignoring quality and service). When a cost leadership strategy is adopted, competitive pricing drives all manufacturing and promotion decisions. To achieve cost leadership, the company may construct more efficient production facilities and install new processing equipment. It practices tight cost and overhead control and vigorously reduces costs. Access to capital, control of labor, increasing efficiency of manufacturing processes, and efficient distribution become the prime success factors in achieving cost leadership.

Some companies emphasize low cost production of undifferentiated (commodity) products. They compete by whipping production into greater efficiency and maintaining competitive prices. Others compete by increasing the value added to products by kiln drying, planing and remanufacturing processes. A focus on servicing a specific segment of the existing market is a workable strategy.

Other companies combine cost leadership and differentiation strategies in a focus strategy aimed at a specific market segment.

2. ***Continue making existing products and add new markets*** New markets may be developed by:

- Segmenting existing markets
- Focusing on serving customers in a section of the market more efficiently or effectively than competitors
- Conducting a market study to discover new markets
- Studying new trends to identify emerging markets
- Searching for niche markets
- Investigating the export market

Further considerations for the existing products/new markets strategy are discussed in Figure I-5.

3. ***Introduce new products to sell to existing markets*** A company ought to view development of new products favorably when its position in the existing market, arrangements with distribution channel members, and name familiarity give the new product the advantage of piggy-backing on the existing product line.

Plywood producers build waferboard, or medium density fiberboard (MDF) or other composite panel plants to reap the advantages of selling to existing markets with which they are already familiar. RCA became an early leader in color televisions by utilizing its existing distribution system to reach the same market to which it formerly sold black and white television sets.

4. ***Introduce new products to sell to new markets*** Adopt a total diversification program. Seek new markets with room for growth and a high potential for adding value.

SELECTING A STRATEGY

In the 1970s and 1980s, American manufacturers of basic products—lumber, aluminum, and textiles, among them—faced the challenges of foreign competition by adopting differentiation strategies that changed the focus of their organizations. Some scrapped their commodity businesses and moved to high end products. Some diversified by latching onto high growth businesses. The resulting conglomerates bore little resemblance to the original companies; some even changed their names: U.S. Steel became USX; American Can became Primerica.

Others followed a different drummer, putting their effort into making products they knew how to make, resulting in greater efficiency, better quality, and lower costs to compete with imports.

FIGURE I-5
Considerations in Using the Existing Products/New Markets Strategy

Understanding the expectations of potential customers unlocks the door to success with this strategy.

Both product differentiation and cost leadership strategy methods may be used, so long as neither one is compromised.

> A manufacturer producing kits for do-it-yourself outdoor decks llustrates the potential conflict between strategies. The company markets a complete kit for a do-it-yourself outdoor deck. The kit is a quality package, including instructions, required hardware, and virtually all material necessary to construct the deck. The package features wider than usual deck boards, the latest CCA treatment, and the most durable hot-dipped coated nails. The quality kit retails for a higher price than other packages. The promotion differentiates the product on quality—the wider boards, non-corrosive nails, preservative treatment, and completeness of the kit. It is positioned as a quality product for consumers who want the best.
>
> To expand market share, company management decides to also produce a less expensive package, using lower grade wood and common zinc coated nails. The promotion does not sufficiently differentiate between the quality kit and the lower cost kit. Soon purchasers of the less expensive package kit begin to complain of early deck failure (caused by the corrosion of the common zinc coated nails). The customers have not clearly understood the difference between the two kits. The resulting complaints blemish the manufacturer's hard won reputation for quality.
>
> In this example, the lower price strategy proved incompatible with the product differential strategy.

Developing new markets for existing products is also useful to extend the mature stage of a product lifecycle.

> A model for extending the mature stage is the marketing of baking soda. For many years, homemakers bought baking soda only for baking. To expand its market, Arm & Hammer Company utilized the odor adsorbing characteristic of baking soda and repackaged the baking product as a rug deodorizer and refrigerator deodorizer. Arm & Hammer encourages continuous sales by printing a calendar on each package to remind users to replace the box in the refrigerator every three months.
>
> The mature stage of the lumber life cycle is extended by shrink-wrapping a surfaced and stained board cut-to-size, and including brackets and screws in the package. The board can then be used by a new market—the do-it-yourself householder who is all thumbs or has no woodworking equipment at hand—to install a necessary shelf. Adding an attractive label with instructions on installing the shelf completes the package.

The emerging global markets heighten the dilemma about the marketing direction a company might choose. Developing Asian countries play musical chairs as markets in the Pacific Rim grow. The 1992 European Economic Community creation of a huge single market from many national markets is also becoming a major force in marketing. Both large, well financed companies and smaller companies are responding to the globalization of markets as their resources permit.

A strategy idea that has gained favor with many large companies is the identification of the "driving force," the primary selection factor that

FIGURE I-6
The "Driving Force" Strategy

The majority of management decisions in operating a business are about four factors:

1) The products made
2) The markets served
3) Manufacturing and marketing capabilities
4) The results desired

The framework that guides management decisions on these factors is called the "driving force," defined by Benjamin Tregoe and John Zimmerman as "the primary determiner of the scope of future products and markets." Tregoe, Chief Executive Officer of Kepner-Tregoe, Inc., an international management consulting firm for Fortune 500 clients, and Zimmerman, Senior Vice President of the company, have identified nine forces that "drive" or guide successful companies:

1. Products offered
2. Market needs
3. Technology
4. Production capability
5. Method of distribution
6. Natural resources
7. Method of sale
8. Size and growth
9. Return and profit

All nine forces are at work in every company, but experience indicates, according to Tregoe/Zimmerman, that the most successful companies are guided mainly by one and only one of these major driving forces.

Source: Benjamin B. Tregoe and John W. Zimmerman. "Top Management Strategy", published by Simon and Schuster, New York, 1980.

FIGURE I-7
The Company that Did Not Know Where It Wanted to Go

A wood products manufacturer (we will call Mixedwood Lumber Company) like many wood products companies, was started by a free-wheeling entrepreneur, and continued to grow by producing quality products and serving its market well.

The desire to continue an upward growth curve and a good financial statement led the firm to add a new plant. Close to raw materials and efficiently equipped, the new plant began operation when the market was brisk and quickly achieved a positive cash flow. Buoyed by the success of this purchase, the firm felt confident that it would continue to expand. The banks readily granted short term loans, which Mixedwood used to buy five additional plants that produced diverse construction materials, many not related to the wood industry.

The investments were predicated on these plants continuing to run at capacity. When the lumber market declined, capacity production could not be maintained. Mixedwood then diversified into new and different lower priced product lines in markets with which it had little experience. The lower priced product lines did not capture sufficient market share to recoup the investment, and worse, they eroded the firm's reputation for quality.

Without a clear sense of direction, top management decisions were erratic and costly. The original entrepreneur lost his company to a larger firm. It took the new management team several years to return Mixedwood to a sound financial position. The new management accomplished the turn around by focusing on specialized wood-based products they knew how to manufacture and sell. They concentrated on improving and extending the product line to serve their existing markets.

determines the scope of products offered and markets sought. The driving force concept is described in Figure I-6.

All strategies are dynamic. They change with the competitive environment. Companies revise their driving force as economic conditions and their competitive positions evolve. New top management or new production capabilities may also lead to reconsideration of objectives.

Many nations are selecting a new driving force on utilization of their timber resources. In 1985, Indonesia switched from a natural resource driving force, in which exploitation of its timber resource was the basis for policy, to a products offered strategy. The new strategy prohibits the export of logs and has fostered a log processing industry to make and export products, rather than the raw material. In the late 1980s, British Columbia adopted a products strategy to add value to lumber.

Selecting A Marketing Strategy Is An Essential Management Function

Is pinning a label on a company's strategy an academic exercise? Does having a strategy make a difference? Most business leaders believe that a company with a clear idea of the direction it intends to go can avoid many mistakes.

The consequences when management does not have a clear idea of the direction it wants to go can be disastrous. Figure I-7 relates the sad tale of a company that did not know where it intended to go.

Frequently, larger firms devote time and energy to devising strategies but they don't seem to "pull it all together." One of the larger wood processing companies that prefers to remain anonymous adopted the corporate strategy of becoming the leader in the wooden door market in the United States and Western Canada. But the division that manufactured the doors had a different objective. It decided to diversify into a new non-wood product and put its efforts into penetrating the tubular furniture market. Result: chaos. This inner conflict may not be as likely to happen in a smaller company.

MAKING MARKETING DECISIONS IN A CHANGING ENVIRONMENT

Current conventional wisdom holds that wood-based companies face five major changes as the 20th century closes. Each change has the potential to force management to reevaluate product offerings and rethink its place in the market:

- ❑ Dislocating fluctuations in the economic cycle.
- ❑ Increased competition nationally, internationally, and from substitute materials.
- ❑ A decline in the ready availability of the wood raw material.
- ❑ Demographic shifts in the population mix.
- ❑ Changes in lifestyles and life cycles.

Marketing pundits have spent hundreds of hours devising methods to guide management decisions in a constantly changing environment. One of these methods is the product portfolio analysis developed by The Boston Consulting Group. Using terms that have become standard business shoptalk in market analysis, The Boston Consulting Group matrix analysis (shown in Figure I-8) is based on the assumption that cash flow and profitability are closely related to sales volume.

A product portfolio analysis is a five step process:

- Data acquisition
- Financial analysis
- Product classification
- Analysis of cash flow/profitability/sales volume relationships
- Marketing decision

FIGURE I-8
Product Portfolio Analysis (Boston Consulting Group)

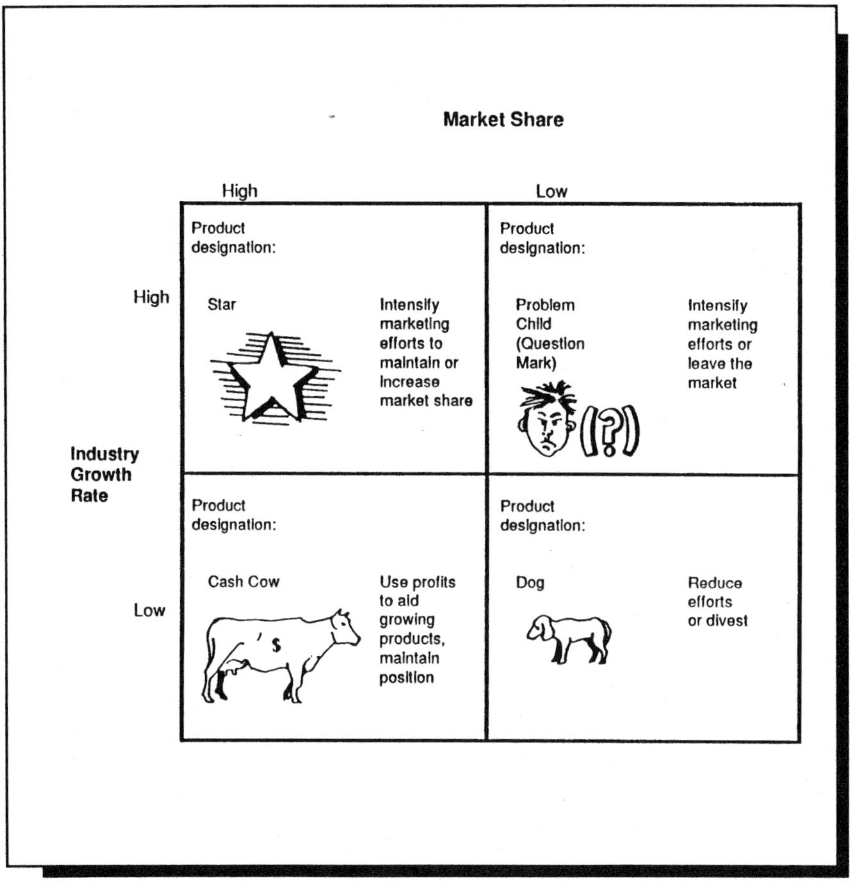

Data acquisition calls for gathering data on:

- ❑ Market share—the company's share of the total market for the product
- ❑ The reported annual growth rate of the market

Financial analysis requires determining profitability of the product and studying cash flow.

Many companies do not analyze profitability by product; they do not know their market share; they are unable to determine whether a decline in sales can be attributed to external factors—a low growth market, for instance—or internal factors, such as cash management. Acquiring the data in itself may help management make more timely, wiser marketing decisions.

Product classification according to the Boston Marketing Group code:

In a product portfolio analysis, products are classified as:

- ❑ cash cows—products with a high sales volume, probably in the maturity stage of the product life cycle, but in a low growth market. Cash cows are profitable products that earn cash for the company.
- ❑ stars—products that are capturing a high share of a rapidly growing market, frequently in the growth stage of the life cycle.
- ❑ dogs—products that are not doing well compared to competitive products. Dogs usually are in a market with little prospect for growth, probably near the end of the product life cycle.
- ❑ problem children, also called question marks—products with a low sales volume in a fast growing market.

Analysis of cash flow/profitability/sales volume relationships means formulating the data acquired into a graphic demonstration of the product position in the market place.

The product portfolio analysis delivers its own package of benefits to managers, especially as a guide to decisions on handling faltering products, investing cash in expansion or redesign, increasing promotion, adjusting prices, and other situations that arise in marketing.

Is the product profitable, with a high sales volume in a static market? The company may milk this cash cow for funds to develop new products or nourish it by improving operating efficiency to maintain the cash flow. Sawmills may find a cash cow in a large volume of lumber sales. While one sawmill may use the profits from selling lumber to develop secondary products, another may opt to protect the cash cow by investing in equipment to improve operating efficiency.

Is the sales volume of the product low in a rapidly growing market? The product is a question mark that may consume large amounts of cash if the company decides to increase its share of the market. When faced with a question mark product, a company chooses between intensifying marketing efforts to elevate it to a star, or deleting the product.

Is the product profitable; has it captured a high share of the market, and is the market growing rapidly? If so, that product is a star! Maintaining a favorable competitive position may consume cash reserves. The star of a cabinet manufacturer may be a line of frameless cabinets. The success of these cabinets encourages competition. How much cash should management commit to maintain the competitive position of the frameless cabinets until the star becomes a cash cow.

Is the sales volume of the product sluggish in a market that promises little growth? The product is a dog that needs to be leashed. Should the company invest in additional promotion to increase market share, or will that be throwing good money after bad? Perhaps a redesign of the dog would appeal to a specific segment of the market. A manufacturer of wood furniture for the baby market finds its product line becomes a dog when the birth rate declines. To salvage remaining sales potential, the furniture might be redesigned to appeal to the more affluent end of the baby population. (See the section Niche Markets in Chapter IV, "Markets—Finding Segments, Niches, and Trends.")

MEGAMARKETING

Marketing viewed as the development and distribution of products and services that create and satisfy customers sometimes runs into an apparently insurmountable roadblock. An unforeseen barrier separates the marketer from potential customers. Megamarketing was devised to handle that contingency. In megamarketing, the marketer finds it necessary to serve the wants of non-customers in order to gain the privilege of serving customers.

Two circumstances are moving megamarketing into the realm of everyday operations: One is the increasing interest in entering international markets. The second is the increasing impact of the American public on the availability of timber resources and the harvesting and manufacturing practices of the forest industry.

> An example of the first circumstance is marketing to Japan. Many affluent Japanese executives are quite partial to American type "stick built" houses. Entire developments of American style homes have been successfully completed in Japan. However, visible and invisible barriers shelter many Japanese markets and institutions from the inroads of American wood products. The Japanese government participated in the American housing developments, but unions and a complex system of distribution have slowed the acceptance of stick built homes. The Western Wood Products Association (WWPA) and the American Plywood Association (APA) have

been involved in megamarketing to Japanese institutions to overcome the barriers to American type building.

An example of the increasing impact of the non-customer public on the forest industry is the construction of a new plant in Minnesota. When MacMillan Bloedel announced its intention to construct a plant to manufacture a new composite panel in Minnesota, environmental and other groups objected with claims that the new facility's quest for wood resources would "rape" the local forests. MacMillan Bloedel's decision to locate in Minnesota was a victory for the energetic Minnesota Department of Economic Development and the Department of Natural Resources. It appeared for a while that litigation by the Sierra Club, et al, against the Minnesota Pollution Control Agency and the Department of Natural Resources would thwart the project. Finally, the parties reached an agreement, which included provisions for a generic environmental impact statement on the impact of harvesting that could affect the entire state.

In each example, the support of non-customers had to be won before customers could be served.

Megamarketing involves both political skill and application of political power. Megamarketing, a term devised by Philip Kotler, Professor of Marketing at Northwestern University, who is well-known in marketing circles, adds two more *P's*—political power and public relations—to the four traditional areas of marketing. For markets blocked by governments, labor unions, and public interest groups of all types, marketing skills with product, price, place, and promotion are not enough. The megamarketing concept holds that a megamarketing strategy should be enlisted to win the support of non-customers before marketing attempts to create and satisfy customers.

Marketing traditionally involves consumers, distributors, dealers, and industrial customers. Megamarketing deals with additional players: legislators, labor unions, government agencies, and public activists promoting myriads of special interests, from preserving scenic areas to protecting spotted owls. Megamarketing takes much longer and requires larger investments than marketing products or services. Megamarketing involves more than economic value: it requires political maneuvering.

Few wood products manufacturers have had to develop political skills. Even if they have the skills, it is highly unlikely that they would have the time or money for megamarketing. Yet, megamarketing affects their ability to export.

The victory of Pepsi®-Cola in gaining entry to the 730 million people market of India illustrates megamarketing. Coca-Cola had withdrawn from the market in 1978 as a protest against Indian government policies. When The Coca-Cola Company tried to reenter later, it found that Pepsico had already effectively used megamarketing to strike a deal with India. Pepsico's effort to form a joint venture with an Indian group helped win government approval despite considerable domestic opposition. Pepsico's agreement to assist India in exporting agro-based products, and its promise to reach rural areas as well as the urban markets was persuasive. Moreover, Pepsico offered to bring to India new food processing techniques, along with packaging and water treatment technology.

Barriers to international markets may include political favoritism, as well as tough negotiators, discriminatory tariffs, protectionist actions, cartel agreements, unfriendly distribution channels, demands for burdensome concessions, and outright refusals to cooperate. Juggling these balls is a formidable challenge, one that few individual wood industry companies have the resources to undertake.

The trade associations in the wood industry are practicing megamarketing, attempting to tear down the walls so the product and service marketing programs of individual companies can succeed. The international trade organizations in individual states and the federal government are all working to remove barriers that may make it difficult for private companies to do business in other nations.

THE TOOLS OF THE MEGAMARKETER

The two principal tools of the megamarketer are power and public relations. Power is a "push" tactic that uses incentives to win the support of influential organizations and government officials. Public relations is the "pull" maneuver that helps open the doors by cultivating public goodwill.

The power tactic means identifying the people and institutions to be influenced. Who has the power to say "yes" or "no"? What are the motives and ambitions of the key persons in the local power structure? How can they be induced to cooperate? In some countries, the inducement may be a surreptitious payoff or contributions to political campaigns. In other areas, it may be enhancing the prestige of powerful persons, or providing pleasure with entertainment, or paying for travel.

Public relations is effective in gaining public support, but it should not be expected to miraculously turn a disaster from a production accident or explosion that injures hundreds of people into a favorable public perception. Organizations are finding that good citizenship benefits the company as well as the community in which it is located.

MEGAMARKETING TO GAIN PUBLIC SUPPORT FOR THE FOREST INDUSTRY

Although the aim of megamarketing is to remove barriers to markets, megamarketing strategies can be applied to lowering the friction between the forest industry and many activist public groups. Between

1986 and 1989, more than 30 percent of media coverage of named corporations was negative. It was probably higher for many in the forest industry. Public relations specialists say four positive or neutral units of news coverage are required to equal one negative unit. Gaining public support translates into benefits for both the forest industry and the public.

Megamarketing employs political power. The forest industry can develop that power by:

- ❑ Searching for groups and organizations that have the same stake in forestry and forest products as the forest industry. In addition to the usual Chamber of Commerce, civic development groups, and economic developers, the large, growing activist groups for the homeless are using their muscle to push affordable housing, a major user of forest products. Allies may include senior citizens, young families, hunters, and other non-timber users of the forest.
- ❑ Organizing those allies into a political and educational coalition. Collective power is more effective than individual action.
- ❑ Cultivating organizations and individuals that do not have a direct stake in the action and may be otherwise indifferent (but also have a voice in the final decision).

The "pull" strategy of megamarketing is public relations. Employ public relations to gain public support by:

- ❑ Recognizing public concerns and ameliorating any legitimate problems for the community caused by harvesting or manufacturing forest products. Establishment of buffer zones, repair of damaged sites, constructing roads to ease traffic flow, compensation for damages have all been used to ease perceived or real burdens.
- ❑ Determining the organizations and institutions that may be influenced by lobbying, advertising, legal action, and other public relations activities.
- ❑ Formulating the message to be delivered to the public, adopting a clear strategy for action, and implementing the strategy with tested public relations tactics.

II

WOOD PRODUCTS

. . . something for everyone

CHAPTER SUMMARY *The purpose of a product is to satisfy the customer's expectations and provide solutions to customer's problems. Without products or service, marketing has little purpose. Without marketing, products drift aimlessly. This Chapter on marketing wood products examines how each product meets buyer's expectations and looks at expansion opportunities for the wood industry.*

Singularization, a concept that differentiates wood products from products made of competitive materials, is introduced as a marketing tool. The experiences of wood products producers throughout the world show the potential for the development of new wood products and expansion of product lines. Guidelines for marketing four types of innovative wood products are presented, with options for marketing new and established wood products.

WOOD PRODUCTS

Everyone uses wood products. Children play with wooden toys; wood furniture is used at home, in offices, in institutions. Homes and apartments use wood in construction and cabinets. From wood are fashioned bowls and mousetraps and hundreds of decorative and useful objects. Wood components are used by industries in the manufacture of industrial and consumer goods. Wood products are so much a part of everyday life most people give little thought to how they are made or what they are made of.

Over the years, consumers have purchased items made of materials other than wood: metal kitchen cabinets, brass and glass furniture, carpeted floors, wallboard, cement, and plastics. But, though they sometimes lose market shares with each substitute, wood products continue to rebound in public preference.

The increasing cost and decreasing availability of timber, global competition, and dramatic changes in the needs and wants of consumers the world over are altering the character and practices of the forest industry. Many forest industry companies that previously limited production to primary processing of commodities are now shifting to the manufacture of secondary wood products. This shifting is changing the way companies manage sales. Products need to be marketed. Marketing demands a mind set that is quite different from selling commodities.

WOOD PRODUCTS VIEWED AS A PACKAGE OF BENEFITS

In marketing, a product is considered a package of benefits, rather than a specific item to hold, feel, or see. That perspective explains why wood items sell for higher prices. They provide more benefits in their progress from commodities to finished products. Figure II-1 illustrates the package of benefits concept through a series of concentric circles. In the middle are the commodities—lumber, plywood, particleboard, and composite boards. Commodities are the products of primary processing

FIGURE II-1
Product as a Package of Benefits

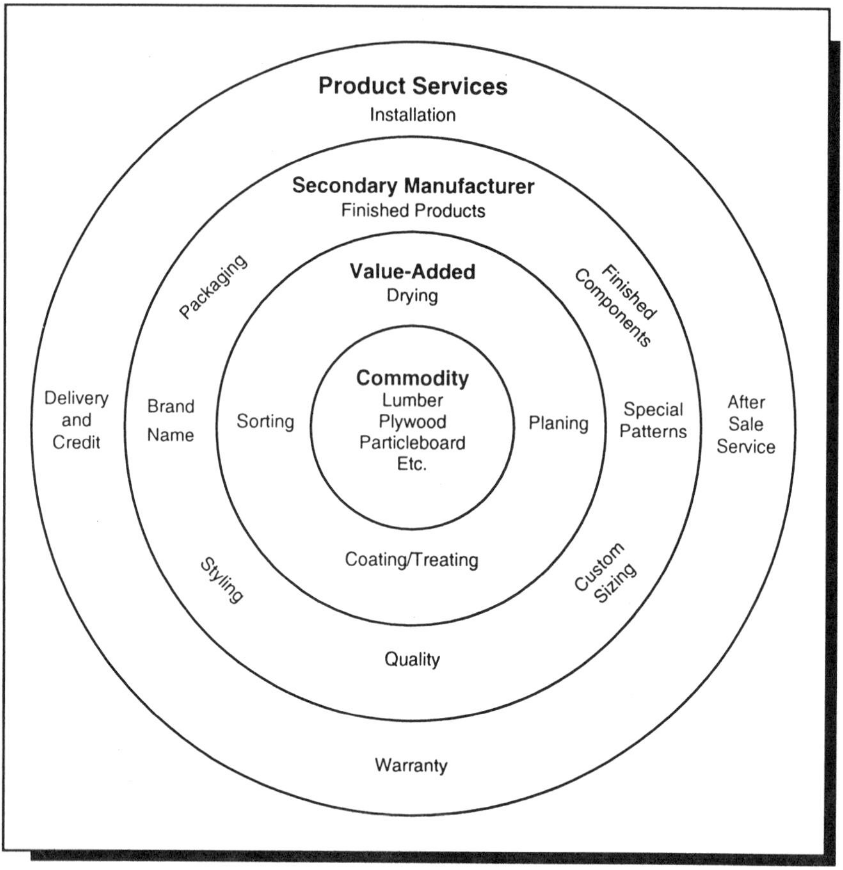

and, on the basis of dollars per thousand board feet, sell for lowest prices.

Some manufacturers add value by sorting, coating/treating, drying, and planing, as shown in the second circle. Value added products command higher prices.

Secondary manufacturers utilize wood commodities to make products that meet specific needs of industries and consumers. Processing an ordinary stud by painting both ends is one step that adds to the sales value. Grading out #1 kiln dried studs from the 2 × 4 Standard and Better studs provides a service that nets additional income. Sizing and packaging for the do-it-yourself retail market converts a stud from a commodity into a product that consumers can use without alteration. Additional services facilitate the distribution and use of finished products and components.

As shown in the fourth circle, delivery and credit are product benefits for distributors and retailers. By quoting delivered prices, the stud producer provides a further service to the retailer. Providing mixed loads also enhances the stud's value to retailers.

Two concepts are the foundation for the package-of-benefits marketing philosophy:

- ❑ The closer the product approximates the shape, form, and properties required by the user, the higher the value to the user.
- ❑ The more benefits a consumer perceives that the product provides, the greater the value to the consumer.

Take installation (shown in the outer circle of Figure II-1). The installation benefit is the basis for packaged kits for the do-it-yourself market. Pre-cut kits for constructing decks illustrate. With homeowners building close to a million decks a year, manufacturers are busily devising new products to make deck installation easier and to improve deck appearance and durability.

> Deck-Tie board and rail connectors introduced by Simpson Strong-Tie® Co., San Leandro, California, eliminate surface nails.
>
> New post and joist connectors make construction easier.
>
> Factory assembled connector packages that are included in the deck price are gaining popularity.
>
> Pre-cut 10-feet-by-10-feet and 14-feet-by-14-feet deck kits with all lumber, hardware, and nails included in a packaged product make construction easier and faster, and result in improved appearance and life of the deck.
>
> Doyle Simpson, National Accounts Manager of Simpson Strong-Tie®, explains the marketing strategy. "To keep (our deck business) growing," he

says, "we've opted to go the project route instead of the product path. We're looking first at yard projects that can be prepackaged and then designing products to make them feasible."

Chris Taelman, editor of the trade magazine *Do-It-Yourself Retailing*, agrees. "There is more emphasis in the retail marketplace on project sales, rather than just product sales," he says. "Both consumers and retailers benefit from bundling items in project sized packages."

The wood cabinet industry considers value the key to continued cabinet market growth and is expanding the package of benefits offered to consumers. The benefits extend to adding accessories that make the cabinets more convenient for modern users. Roll-out trays, tilt-out shelves, adjustable shelves, wine racks, spinning shelves, trash can holders and special hardware which used to be limited to semicustom cabinets are now included in stock cabinets. Accessories help satisfy the high value expectations of do-it-yourself customers who represent 25 percent of all kitchen and bath remodeling.

The German cabinet industry includes these benefits as standard items in its cabinet and furniture products.

This package-of-benefits concept applies equally to industrial wood products. Illustrations in Chapter V on product differentiation and positioning demonstrate that an industrial customer benefits from the package approach by saving money and time.

A MARKETING PERSPECTIVE OF FOREST PRODUCTS

A fundamental difference between selling and marketing is how products are viewed. Lumber as a commodity is characterized by its physical properties and dimensions and sold by grade: a 2 × 4 from one manufacturer may be indistinguishable from a 2 × 4 made by another processor. A shipment of 4/4 red oak from one producer may not be separable from that of another producer. When the lumber is shipped out of the plant the producer often does not know where or how it will be used.

In marketing forest products, who the customer is and what the customer will do with the product is a major consideration. A manufacturer who identifies a customer's needs is marketing, rather than selling.

The first step in marketing is examining the wood product or products from a marketing perspective. This means viewing the product from every angle: use; species; life cycle; and sensitivity to economic, demographic, and lifestyle factors.

Examining the product from every angle is critical to:

- ❑ Determining market growth opportunities
- ❑ Obtaining information about potential customers
- ❑ Identifying competitors
- ❑ Selecting a marketing strategy
- ❑ Finding the best distribution method to move the product out of the factory and into the hands of customers
- ❑ Choosing the most effective method for promoting the product.
- ❑ Evaluating the importance of packaging for the product
- ❑ Calculating the price of the product
- ❑ Obtaining capital to move into the manufacture of a new product line. (Financial institutions now take as hard a look at the marketing strategy as at the product itself.)

Wood species may influence product differentiation and positioning. If a product is closely identified with a particular species, substituting another species may introduce marketing problems. If the product is interior flooring or tool handles, is there a marketing problem in substituting for oak flooring or hickory handles? Does the substituted species add another marketing challenge for the manufacturer?

Product Lifecycle

A product's position in the product lifecycle influences the selection of a marketing strategy. Every product goes through a lifecycle in its sales history, as shown in Figure II-2. The product lifecycle goes through four stages:

1. Introductory: The product is new. When introducing a new product the manufacturer typically faces an initial low sales growth rate and low cash flow.
2. Growth: The product has won customer acceptance and the sales volume is growing rapidly. At this stage, unit price levels tend to be high. For example, packaged kits for use by do-it-yourself consumers are still in the growth stage of the product lifecycle. As another example, the ski industry was in the high growth stage during the 1970s, with the number of skiers growing at double digit rates.
3. Mature: The product has been around for a while. It has reached its highest sales peak. At this stage, profit margins may begin to decline from those experienced in the growth stage. For instance, traditional wall wood paneling may now be in the mature stage.

FIGURE II-2
Stages in a Product Lifecycle

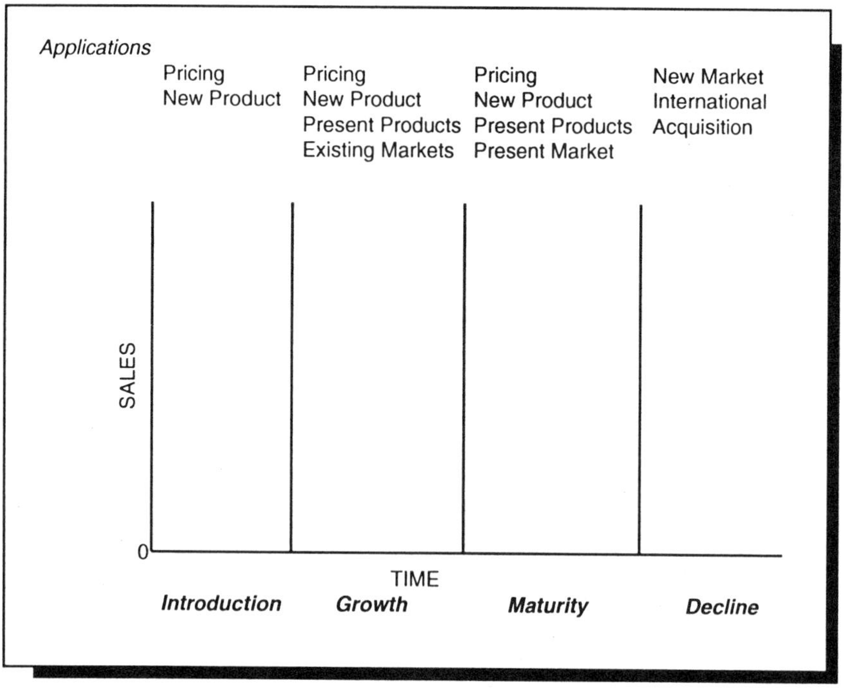

Products sometimes enter a second and third cycle during the mature phase, extending this phase almost indefinitely. Hardwood flooring has experienced a second cycle of its mature phase as a result of new finishing techniques that minimize maintenance. Windows might be considered in the mature stage, but more builders are using styled windows to distinguish their houses, offering window manufacturers opportunities to extend the mature stage. For example, Marvin windows created the Corner Window to let builders create glass corners without posts or joints. The Norwegian H-windows are also extending the mature cycle of wood windows.

4. Declining: A change in consumer preference, a shift in the population mix, competition, the introduction of new substitute materials, or other factors cause declining unit and dollar sales and margins. Wood furniture for exterior use experienced a decline when "resin" outdoor furniture was introduced. ("Resin" is the furniture trade designation for plastic furniture using resin).

Sensitivity to conditions over which the manufacturer has no control affects marketing strategy. Many mills have a high degree of sensitivity to housing starts because when housing starts decline, it usually spells hard times. The growth of the home repair business helps smooth out the ups-and-downs of the construction cycle. Repair and remodeling sales have helped fuel the growth of secondary processing.

Wood office furniture markets are sensitive to the business cycle. The industrial markets for furniture components and hardwood dimension stock are derived from the consumer furniture market. If the product is pallets, the pallet demand fluctuates with industrial activity. Demand for fencing, patio furniture, and similar products is typically seasonal. Demand for wood gift novelties increases with the Christmas season, and locally may also increase with the tourist season.

Other facets of the economy that affect sales are inflation, interest rates, unemployment, and the gross national product (GNP). Inflation diminishes real income, leaving less money for discretionary purchases of wood products. High unemployment also discourages spending on discretionary items. The GNP measures the annual volume of goods and services produced in the United States, and is often used as a measure of anticipated demand for products.

Demographics and *lifestyle*s influence sales of wood products. The do-it-yourself decks are sensitive to the aging of the population (increasing as the baby boomers reach middle age) and to lifestyle changes such as "cocooning"—a term that describes the trend toward refocusing on more leisure time at home. Lifestyle experts suggest that families are willing to invest in making their homes more livable as an option to spending time and money on entertainment away from home. Early retirements and increased leisure mean opportunities for do-it-yourself home improvement projects. A baby boom or boomlet means a growth in schools and school furniture.

The broadened perspective acquired from the New Product Checklist, Figure II-3, can help the marketer discover potential marketing mistakes before they happen. With this information at hand, a company can avoid having only products in the mature stage of their product lifecycles, targeting an unsuitable market, or choosing a less effective distribution channel. It also serves as a guide for selecting product differentials, promotion, and price strategy.

MARKETING BREEDS OPPORTUNITIES

Marketing is a never-ending task, requiring constant updating, checking, and alertness. Most smaller businesses pile new product development on top of the many tasks already filling the manager's or boss's day. Given the work, money, and time requirements, what motivates a sawmill to expand into value-added processing, or a hardwood

FIGURE II-3
New Product Checklist

1. What does the company hope to accomplish with this new product?

   Increase profits___ Diversify product lines___

   Stimulate sales of companion or accessory items___

2. Who would purchase this product?

   Do-it-yourselfers___ Owners of new homes___

   Upper middle income families___ Women___

   Remodelers of older homes___ Others________________

   Manufacturers of panelized or modular prefabricated housing___

3. What product features are of most concern to these consumers?

   Ease of installation___ Weight___

   Pattern___ Unique status item___

   Durability___ Safety___

4. Do competitive products respond to these concerns?

   Similar products made by other companies___

   Products made of other materials—wallpaper, plywood, laminated composite board, plastics___

   High priced products___

5. Which competitors have a large share of the market?

   Single competitor___ Several competitors___

   Does any competitor have a commanding position in the market___

6. How large is the present market?

   Present size___ Growth rate___

   How much does the company expect to sell each year in dollars?___ In units?___

7. Can product differentiation be an effective strategy?

   Is the product sufficiently different from others?___

   Do the product features stand out from competition?___

8. Will buyers consider the new product superior to similar products they are now using?___

9. Will it be possible to position the product favorably?___

10. Is this the best time to introduce the product?

    Appropriate season?___ Unemployment increasing?___

    Homeholders hesitant about improving homes?___

FIGURE II-3
New Product Checklist (Continued)

11. Will the purchaser need to learn new techniques to use the product?
    Purchase new tools?___ Learn new habits?___
12. Do the physical characteristics of the product limit use?
    Products of a size and weight that women can handle?___
    Will older persons be able to handle it without assistance?___
13. How long to develop and bring the product to market?___
    Is the time frame within set company objectives?___
14. Is the product
    A shopping item___ A prestige purchase___
    Unique___ Appropriate as a gift___

mill to expand into furniture dimension, or a company to produce a butcher block cutting board to sell through gift stores?

Opportunity! Opportunity to influence sales, opportunity to be innovative and creative, opportunity to gain some control and escape from the dictatorship of many economic factors over which an individual has little control. Marketing forest products breeds opportunities that commodity producers often miss.

Opportunities Abound

Outdoor furniture, known as casual furniture in the trade, has evolved from the utilitarian, sturdy look to high-style, with elegant frames, new foam, water resistant cushions, brilliant colors. This trend calls for the makers of redwood and cedar furniture to review and update the design of outdoor furniture.

The population is getting older and many avid gardeners can no longer get down on their knees to garden. This creates opportunities for manufacturers of wooden containers, especially those mounted on wheels, and for accessories that ease the labor of gardening! The growing interest in do-it-yourself projects has created opportunities for packaging projects similar to deck packages.

Who would have thought that thieves would create opportunities for wood products? So many thieves stole aluminum highway signs, bridge railing, and other metals for sale to recyclers in Tacoma, Washington that the Highway Department is replacing aluminum tubes with wooden railings. Likewise, the interest in country furniture opened opportunities for manufacturers of country style furniture and accessories. The renaissance of wood furniture with hand painted detailing opened up opportunities for small and medium sized craft shops.

Hundreds of wood products are marketed. Figure II-4 shows only a few of the more familiar wood products on the market today.

In addition to opportunities for sales and profits, other reasons motivate wood products businesses to investigate new products and new markets:

- ❑ Diversifying product lines
- ❑ Leveling seasonal sales fluctuations
- ❑ Expanding sales
- ❑ Increasing profitability
- ❑ Utilizing existing equipment
- ❑ Retaining present employees
- ❑ Replacing products that are reaching the decline stage in the product lifecycle
- ❑ Replacing markets captured by foreign competition
- ❑ Replacing products made obsolete by new technology

FIGURE II-4
Wood Products Are Used Everywhere

Chairs: dining room, rocking, easy, children's . . . tables: high, low, end, side, night, library, coffee, tilting, sewing, trestle . . . entertainment centers . . . sideboards . . . footstools . . . high stools . . . garden furniture . . . high chairs . . . cribs . . . bunk beds . . . hide-a-beds . . . ladders . . . carts . . . bookcases . . . cabinets: corner, vanity . . . computer work stations . . . beds. . . mirror frames . . . picture frames . . . bins . . . chests: blanket, cedar. . . chests of drawers . . . files . . . wagons . . . swings . . . cabinets . . . decks . . . desks: rolltop, computer, writing, slant front . . . boxes: jewel, key . . . armoires . . . lamps . . . fences . . . sleighs . . . houses: play, tree. . . rocking horses . . . floors . . . trusses . . . roofs . . . mantles. . . windows . . . doors . . . solariums . . . walls . . . mouldings . . . ceilings . . . partitions . . . pews . . . dolls . . . shoes . . . mousetraps . . . handles . . . bread boards . . . salt and pepper shakers . . . cooking tools . . . clocks . . . stairs: spiral . . . homes: stick-built, manufactured, log . . . do-it-yourself construction . . . ready-to-assemble furniture . . . decks, porches, patios . . . hot tubs . . . pieces for home craftsmen

- Adding new product lines made possible by new technology
- Responding to changes in lifestyles
- Responding to changes necessitated by government regulations
- Adapting to changes in available timber or lumber resource
- Adapting to changes in quality of available raw materials.

WHY WOOD PRODUCTS?

Wood readily qualifies as the ideal raw material for many needs. Smaller companies can turn out attractive, useful items with a relatively small investment. Many companies have started in a basement workshop or garage. Manufacturers of wood items can often find a market niche. Larger companies can improve production efficiency by adding modern equipment. Wood has character, connotes quality, is a renewable resource, and has been a preferred material throughout history.

As consumers know, wood has its downside too. It is subject to attack by insects and fire. It may also decay and warp.

To a producer, wood suffers the disadvantages of being a nonhomogeneous raw material; every species requires processing according to its own physical characteristics. The drying cycle for one species may not work for another. Even for the same species, timber growing in one area may have different characteristics from timber originating in another area.

Silvicultural treatment results in a difference in physical characteristics of wood and these treatments vary with the ownership and type of timber.

Many species have unique wood scents that may influence product and species preference.

SINGULARIZATION—BUILT-IN PRODUCT DIFFERENTIAL

These same disadvantages can be turned into a marketing advantage. Wood comes complete with an exceptional marketing asset—a built-in *product differential,* a marketing term for product benefits that win customer preference over competing products. The forest industry often fails to recognize this built-in product differential. Each wood species can be turned into products that are unique, exotic, and encourage pride of ownership. Using wood characteristics as a product differential is an opportunity offered by singularization.

What makes a wood species extraordinary? Consumers perceive wood as exotic if it is different, unique, singular, out of the ordinary, exciting, novel, or rare. An exotic possession is viewed as a more valuable possession, one which generates pride of ownership. Consider the advertisement by a well-known vacation cruise line shown in Figure II-5. The advertisement's boast that the cruise ship's deck is made of *teakwood* demonstrates the high quality of the appointments in its luxury liners. Probably few passengers are familiar with the properties of teakwood, but the advertisement implies that teakwood spells luxury. If they consult a dictionary they'll find that teakwood comes all the way from Thailand, Java, or Burma and is a large East Indian tree with white flowers—a truly exotic specimen.

> The Prudential uses the uniqueness of wood to illustrate the individual tailoring of its retirement policies. Prudential's advertisement features Donald Henness, a sculptor, with the headline ". . . Donald Henness discovers a sculpture in each piece of wood." The ad quotes Henness: "Every piece of wood is a challenge. You have to treat it like an individual."

Many native species have earned a reputation for quality and beauty—oak, cherry, walnut, white pine, ash, birch, among others. But there are so many pieces that are called merely oak or other familiar species that they have lost much of their valued differences. Eighty percent of wood cabinets are made of oak. Without identification of the uniqueness of the oak used in cabinet line, cabinet makers lose one of their most effective wood product differentials.

To differentiate items made from these species, to engender pride, wood products manufacturers can singularize their products by:

❑ Including the origin of the wood in promotional material. The John Deere Catalog for Home, Lawn, and Garden states that its fine

FIGURE II-5
Teakwood Deck Advertisement

. . . AND SOON YOU CAN BE STEPPING ABOARD THE TEAK DECK OF A SPLENDID HOLLAND AMERICA LINE SHIP TO SAVOR BALMY CARIBBEAN SUN AND THE LIFE OF LUXURY . . . AT SPECTACULAR SAVINGS. HAPPY TRAVELING!

Enjoy A Fabulous Vacation
Of A Lifetime

garden furniture ". . . is not made from scarce, endangered trees, rather from the plentiful Shorea tree of Asia. . . . Shorea is a splendid wood for outdoor furniture because it weathers like teak and it's very hard and durable, so strong and weather-resistant that it's widely used for marine construction, even bridges." This is not an ordinary garden set; it has character! Note how John Deere uses the origin to support the superiority of the product.

Whittier Wood Products, manufacturers of ready-to-assemble (RTA) furniture sold all over the United States, identifies its home office furniture as "made of beautiful kiln-dried red alder from the Pacific Northwest."

- ❑ Mentioning unusual weather conditions that give the wood a unique character.
- ❑ Explaining silvicultural practices that make the wood special. For example, the Wooden Shoe Factory makes hundreds of thousands of authentic wooden shoes (***Klompen*** in Dutch, ***Sabots*** in French) and is a popular tourist attraction in Michigan. The technique of singularization used in their literature and advertising creates a solid image: "Craftsmen of Dutch descent produce all the wooden shoes made here at the Wooden Shoe Factory," their brochure states. "They use only logs of poplar and aspen ***cut from trees grown in sandy soil on high ground to prevent the heart of the tree turning brown.***" (emphasis by the authors.)
- ❑ Describing unusual harvesting problems. For example, Black walnut might be described as the preferred wood for fine furniture. Because it is harvested one valuable tree at a time, good quality black walnut is hard to find at reasonable prices.
- ❑ Mentioning unique characteristics. As an example, Hickson Corporation publishes a booklet called "How to Build a Deck," featuring a construction material they term "outdoor wood." This is their brand name for outdoor pressure treated wood. The booklet notes that "lumber is a beautiful, natural material. And it's renewable." The text explains that outdoor wood is pressure treated, high quality lumber and is a special type of "Wolmanized" wood.

Even apparently unattractive characteristics can provide a benefit. For instance, a pine shelf packaged in shrink wrap available in home center stores includes brackets and instructions for installation. However, a close inspection reveals that the board has slight defects. That does not daunt the manufacturer, who explains on the printed label that the slight defects in the board give the shelf its unique character.

- ❑ Describing how the wood item is made can also differentiate a product. The Wooden Shoe Factory's brochures describes in considerable detail the care and skill used in making the shoes: "The roughing machine, built in 1908. . . . The 1914 vintage boring machine, imported from France, bores out the front part of the shoe. . . . The heeler prepares the shoe for hand finishing."
- ❑ Pulling in any other factor that may make the wood specially interesting. The Eddie Bauer catalog describes its Turned Rail Bed as: "crafted from western lodgepole pine—so named because Native Americans used this sturdy wood for tepees and lodges. Its rugged texture is softened with turned finials, and polyurethane enriched Danish oil seals and protects the wood."

Mike Burrill, President of Eugene F. Burrill Lumber Company, in commenting on the conflict between the western timber industry and environmental organizations, stated: "Many people don't relate the lumber, plywood or paper that they use everyday to the tree that is grown."

The forest industry gives the public little help in relating trees to products. When trees are treated as merely raw material, they are viewed as such. When they are treated as unique products, they assume the standing of unique products.

World-wide public concern about depletion of forests has added another element to the concept of singularization: Assurance that use of the wood is environmentally acceptable.

> This End Up Furniture Company, puts out an attractive brochure that describes the uniqueness of the company using terms such as "our originality," "our factory," "our wood," "our stores," "our office," and "our delivery." One section explains: ". . . we no longer use shipping crates to make This End Up furniture. We use something far better. And it's what makes our furniture so attractive, and so indestructible. Solid, kiln-dried southern yellow pine. . . . Yellow pine is heavy, hard, and high in shock resistance. . . . At This End Up, we're not just conscientious about our furniture. We're conscientious about our environment too. America's Central Southern Region, where we buy our wood, has very active reforestation programs. So although we use large quantities of yellow pine in our furniture, we guard against the depletion of that valuable and beautiful natural resource."

> A full two-page advertisement in November 5, 1990 Fortune magazine by the microfiche producer, Anacomp, proclaims that the environment benefits by saving paper when its microfiche is used: "The Fiche . . . can Save 30,000,000 Trees in North America Alone. Now, Doesn't That Make Us All Breathe Easier?" Under a picture of a majestic forest, the ad continues:

"You'll breathe easy when you consider how many trees you're saving with our image management platform. . . ."

Every wood product is unique. Consider oak. An oak board might have the grain of red oak from the hills of western Pennsylvania, or be exceptionally hard and abrasion resistant white oak from eastern Texas processed into hardwood flooring. Chestnut oak from the forests of southeastern Michigan is known for durability. The oak can be enduring genuine northern red oak, not a substitute similar to red oak. It can be white oak from the valleys in Virginia which survived the unusual freeze of 1925. Or it can be red oak from the ridgetops of the Hudson River Valley in New York.

Any of these descriptive embellishments helps the purchaser of new kitchen cabinets buy additional pride with the purchase. "These cabinets aren't just ordinary oak. They are genuine red oak from New York's Hudson Valley," the homeowner might say. Or, "The new hardwood floor is that really hard white oak that comes from Texas".

When a manufacturer of high end special veneers labels them with the names of the species—Agathis, maranti, Podocarpus, Sumatra oak—the purchaser's mind travels to faraway exotic places, but not quite. For all the customer knows, Podocarpus may be a plastic: made-up names of products bombard the consumer every day. How much more effectively might those veneers be marketed if they were identified as "prime Agathis from the forests of Indonesia," or "Sumatra oak, an exceptional wood from Indonesia."

Softwoods are also singular. Eastern white pine might be identified as the most important conifer in the Northeast and Lake States. The historical background of Maine pumpkin pine lends additional interest to the wood. The numerous varieties of ponderosa pine are all identified simply as ponderosa pine. But the pine from New Mexico and Arizona differs in grain and hardness from that grown in Eastern Oregon and Washington. User comments about ponderosa pine indicate preferences for pine from one location or another. Geographical identification of ponderosa pine should lead to greater consumer satisfaction.

Old-growth Douglas fir lumber has evolved into a more valuable species as result of the spotted owl controversy that has rocked the western forest industry. A state forestry newsletter reported that at the height of the spotted owl controversy, a customer came into a lumber retail yard and requested old-growth lumber because he "wanted something special." The media attention to old-growth Douglas fir has generated a marketing opportunity. Properties can be tied to the age of the tree, diameter, or other silvicultural factors that differentiate one piece of Douglas fir from another.

Differentiating wood with defects attracts customers as well. It has also been found to be the more profitable way to use wood considered too low grade for furniture or cabinets. "Character wood" panels in which character marks give each panel a distinctive appearance have also proven popular with consumers. Embellished with interesting titles—Appalachian Rustic Oak, Bright Knotty Pine, Appalachian Colonial Poplar, Peckerwood Montana Blue Stained Pine, the "character wood" panels are retailed in lumber yards and home center stores.

SINGULARIZATION IS THE HALLMARK OF THE MARKETING MENTALITY

The concept of singularization signals the forest industry switch from a selling mentality to a marketing frame of mind. Traditionally, the forest industry has focused on providing uniform commodities that meet grade and size standards for each species. Industrial customers and construction contractors depend on standardization. But the ultimate consumer is more interested in the qualities of the product that relate to reasons for purchase. And the uniformity, grade, and quality control that help manufacturers of the product are not among the attributes consumers seek when they purchase lumber for their weekend project.

The manufacturers of cabinets, prefabricated homes, doors, occasional furniture, and the hundreds of other wood products are caught in the middle. For manufacturing purposes, uniformity and standardization of raw material are preferred; for marketing purposes, product differentiation is an asset.

However, manufacturers can have it both ways. They may find it necessary to alter some purchasing and processing practices. Sawmills that buy logs from loggers they know are generally familiar with the timber source. By the time these logs are processed into lumber, their singularization identity may be lost. The company that purchases lumber from several sources may not be able to trace the origin of the lumber. A change in purchasing practices to allow for singularization would involve following the origin of a log through as it becomes raw material for a wood product.

Recognizing that a customer buys wood cabinets and wood furniture for the intrinsic value of the beauty and warmth of wood, singularization is an added plus in wood products.

Product differentiation through singularization links the origin of the wood to the additional value of the product to the buyer. The linkage is communicated to the consumer through promotion. Singularization begins in the forest and ends with the retailer. A *point of purchase* (POP) tag is the simplest way to inform the consumer that the item is special. The singularization message is carried through in advertising, public relations, display booths, product information publications, assembly manuals and instruction sheets. The identical singularization message is repeated in the packaging, and in a label affixed inconspicuously but permanently to the product (assuming that the customer will eventually lose or mislay any tags or other information that came with the purchase).

Picture a kitchen cabinet on display in a home center store. The point of purchase tag hanging from the knob informs the owner that the cabinet doors are Northern red oak from the rolling hills of western Pennsylvania. Sales information includes a short description of the special qualities of this particular oak. The display signs in the kitchen cabinet department include the words "Northern red oak from the rolling hills of Pennsylvania." The sales clerks have had some basic instruction in the differences between species, varieties of the same species, influence of site, and other factors that make one piece of wood different from the other. They repeat the singularization message during the sales transaction.

Showing friends the new cabinets installed in the kitchen, the customer announces proudly, "These cabinets are made from Northern red oak from Pennsylvania." The singularization mission is accomplished!

This program has none of the hype that many manufacturers dislike and consumers resent; it merely states the facts of interest to the customer.

The benefits of singularization in high-end-quality products are apparent. Singularization also has value for lower-end products, composites such as particleboard, oriented strand board (OSB), flakeboard, and plywood. For instance "This flakeboard is made from Aspen grown in northern Minnesota" differentiates flakeboard made from species in other regions. Or "the top of this table is made from the rare tanoak from the northern coast of California."

> To cap the case for singularization, consider the pasta business. Pasta La Bella, made by the Missouri-based American Italian Pasta Co., capitalizes on singularization of the wheat used in their pasta. The best quality pasta is made from durum wheat, which comes in many varieties. Pasta La Bella uses five specific varieties of North Dakota durum wheat. Pasta makers don't usually distinguish between wheats when they market pasta as a commodity. Pasta La Bella does. It markets a premium pasta that sells for about

twice as much as ordinary pasta. Instead of selling pasta as a commodity, American Italian Pasta Co. has singularized it into a pasta people respect. The product is premium. The premium image is supported by elegant black packaging featuring a picture of a Tuscan hill town in Italy. Pasta La Bella is so singular Italians import it from America!

SINGULARIZATION AND THE GREEN REVOLUTION

Unique exotic woods convey excitement and lead to pride of ownership. Consumer worry about the environment, however, is adding a new criteria to the value of exotic woods. Potentially endangered wood species convey a negative image to a growing number of buyers. The impact of the "green revolution" is just beginning to be felt by the United States wood-based industry, but in Europe the movement against using certain species is strong.

Public opinion polls show that most Americans believe that saving the environment is a high priority. Most consumers do little to support that belief. But influential consumers act upon their beliefs and will not buy items that are reputed to damage the environment. The number of consumers who are practicing "environmentally friendly" behavior is growing rapidly and merits consideration.

Currently, Americans appear to be divided into five environmental types:

- True-Blue Greens—the 11 percent of the total population who are leaders of the environmental movement. True-blues earn more and have more education than most other Americans.
- Greenback Greens—another 11 percent who are willing to pay more for environmentally acceptable products. They earn about the same as the true-blues and are also better educated than the majority, but they are too busy to change their lifestyle to accommodate environmental concerns.
- Sprouts—who constitute 26 percent of the population are also well educated and have high incomes. Sprouts are inclined to adjust their lifestyles to reflect their environmental beliefs. As the "swing group," their views are important to manufacturers and economic developers.
- Grousers—24 percent of the population, with below average income and less education who are too busy to do anything about the environment.

- Basic Browns—28 percent of the population, the group with the lowest incomes and education, believe there is little individuals can do about the environment, so they don't attempt to do much.

As Figure II-6 shows, environmental awareness varies with income, education, sex, and geography. For premium, or high-end items that appeal to the environmental leaders, singularization should emphasize wood species that are grown and harvested under currently acceptable environmental conditions.

DEVELOPING NEW WOOD PRODUCTS, EXPANDING PRODUCT LINES

Sooner or later, a company loses its competitive position if it does not introduce new products or services. Companies in all industries have concluded that to compete globally they must develop innovative products. Writing in the September 24, 1990 issue of *Fortune* magazine, Edmund Faltermayer doesn't mince words about the role of innovative products in his article "Is 'Made in the U.S.A.' Fading Away?" He writes: "America is losing markets and selling off pieces of its economy to pay for its big import binge. The nation just needs to make more things that people want to buy. Repeatedly, American and Canadian companies are exhorted to innovate new products and services as the key to survival."

In the December 1990 issue of *Inc.* magazine, John Case's article on "The Market Makers" concluded that the 1990 *Inc.* 500—America's fastest growing small private companies—were exceptional innovators and marketers. "Other businesses typically think about serving a market," Case reported, "but *Inc.* 500 companies seem to think about creating a market. They set out to do something no one else is doing. . . . Many of the *Inc.* 500 entrepreneurs have a knack for focusing on what customers need, not just what their own company happens to be selling at the time."

> Forest industry companies might borrow an idea from the success of Resource Conservation Services, the 136th company on the 1990 *Inc.* 500 list. This company created a business with over $6 million in annual sales from the waste wood ash dumped by a paper mill. Resources Conservation Services began by turning the ash into fertilizer and soon expanded into profitable commercial recycling, composting, and wastewater treatment. An example that in the wood-based industry one company's waste is another's raw material!

FIGURE II-6
The "Greenest" and "Brownest" in the American Green Revolution

	"True-Blue Greens"	*"Basic Browns"*
Percent of Adult Population	11%	28%
Male	34	55
Female	66	45
Married	69	59
Not Married	30	41
Median Age (years)	44	42
Median Income (thousands)	$32.1	$21.2
With Children (Percent of group with children under age 13)	34	32
RACE		
White	82	82
Black and Other	14	16
REGION (Percent who live in each Region)		
Northeast	31	17
South	18	48
Midwest	27	22
West	24	13
EDUCATION		
High School or less	50	69
College Graduate or some College	50	31
OCCUPATION		
Executive/Professional	25	11
White Collar	18	15
Blue Collar	19	36
POLITICAL/SOCIAL LEANINGS		
Conservative	43	36
Middle of the Road	26	41
Liberal	28	16

Source: American Demographics, February 1991
(Based on Data from The Roper Organization, New York, NY)

Markets and products are interdependent and interfunctional. Markets are fragile; the growth market of today probably will be the declining market of the day after tomorrow. Company stability and growth is achieved by investment in developing new products or expanding production of existing products. Forest industry management has to satisfy the conflicting demands of market instability and desirable production stability.

The availability of raw material is the first consideration in developing new wood products. When feasible, sawmills locate close to the timber resource and secondary products plants locate closer to the markets. If the logs are processed for a captive product, or developed into a higher value product, proximity to the timber resource becomes less important. In today's global economy, logs are shipped long distances: from the United States to Pacific Rim countries, from Chile to the United States, from the USSR to Japan and the United States. Tropical hardwood logs are shipped from Southeast Asia, Africa and South America to European and American furniture manufacturers as well as those in other countries.

Today, many areas of the world that previously shipped logs now ship lumber and it is often planed, dry-kilned, and ready to use in products. As recently as 1980, Indonesia was one of the major log exporters of the world. A total ban on log exports in 1985 and promotional actions by the Indonesian government changed the Indonesia forest industry to one of the world's leading plywood exporters. By 1990 it became the second largest exporter of tropical sawn lumber. Indonesia now manufactures value-added products rather than shipping roundwood for other countries to process for manufacturing.

Whatever the source, the manufacturer of wood products seeks assurance that lumber, or veneer, or plywood, or composite board will be available when required in appropriate quantities and quality.

> A kitchen cabinet manufacturer with growing national sales was acquired by a large kitchen appliance manufacturer. The new owner, unfamiliar with the wood industry, expanded operating facilities before assuring a continuous supply of quality hardwood. Result: predictable losses, although the order file grew.

> The producer of a new veneer product located a new plant based on assurance of the availability of quality veneer logs, only to find after the new plant was completed that the veneer logs were not of the quality he assumed, were much more costly, and not as readily available.

Standard graded and sized lumber may not be necessary. New sources of raw material are available to the innovative producer. When timber resources become scarce, every scrap is valuable.

The objective is to use these scraps to produce products with the highest possible value. High value products may be produced in a number of ways.

- ❑ End pieces left from a sawmill's standard production of sized and graded lumber. These pieces are often surplus to mill requirements yet suitable for a cut-stock operation or hardwood dimension pieces.
- ❑ Providing custom services such as coating or laminating on particleboard or other composite board. Panel board manufacturers provide the raw material.
- ❑ Producing furniture dimension pieces, hardwood flooring for trucks, and other items from lower quality hardwood logs that may not be usable by a sawmill cutting for grade lumber.
- ❑ Manufacturing pallet stock from the lower grades produced by a hardwood grade sawmill.
- ❑ Producing small specialty items like mousetraps and toy parts from sawmill scrap.

Innovation

Innovation is the key to new products. The new product that is "just what the customer ordered" defines innovation. Some examples of innovative products are: a new technology that gives rise to an entirely new family of products, the new application of an old product that solves a problem, a radically new development that makes possible new products beyond the imagination of people—like TV, video recorders, portable phones, wood decks that survive all weather.

Prefabricated homes that look as upscale as site-built houses, outdoor decks cut to size and packaged for the do-it-yourself customer, wood composites engineered for special purposes, laminated surfaces that combine the comfort of wood and the easy care of plastics, finishes that enhance beauty and utility of wood—all these are innovations that open new markets and invite new customers.

Innovations that are driven to satisfy market demand are, by definition, likely to succeed in gaining a competitive edge. There are four types of innovation a manufacturer can consider. These are shown in Figure II-7.

IF YOU CAN'T LICK 'EM

"If you can't lick 'em, join 'em" is the marketing philosophy behind the research that has given rise to new families of products called composites, produced by joining wood with other materials. Composites

FIGURE II-7
Innovation Strategies

1. ***Innovations based on incrementally new manufacturing technology.***

These innovations result from improvements in manufacturing plants. A market-driven investment in new equipment differs from a product-driven investment. In the booming late 1970s, a Douglas fir/hemlock sawmill borrowed heavily to improve operating efficiency, based on the business philosophy that if you make more product you will make more money. The mill planned to tap into the export market but failed to first inquire into the requirements of export customers. Lacking a dry kiln and basic secondary manufacturing plant, they were unable to make the products and sizes required by the export market. The debt burden incurred for the investment did not result in the expected expanded sales, and the mill could not survive the recession of the early 1980s. Had the new manufacturing technology been market-driven instead of product-driven, the sawmill would probably have been able to pull through. Installation of new processing equipment improves internal efficiency and helps to maintain price levels, but it does not necessarily produce a new product.

2. ***Technological breakthrough***

Manufactured homes are a product of a technological-breakthrough innovation that can be used to provide affordable housing. The 32mm method of manufacturing casegoods and cabinets represents a technological breakthrough that led to the frameless style in cabinet making.

3. ***Application innovations***

The use of treated lumber decking boards and the availability of packaged kits for building outdoor decks are application innovations. Roof trusses, floor trusses, and panelized wall sections are other application innovations.

4. ***Radically new products***

Products that offer the customer an entirely new benefit are a marketing manager's dream, although the radically new products may be a production nightmare. It is management's function to maximize the marketing dream and minimize the production nightmare.

have opened up marketing opportunities undreamed of 20 years ago. Some forest products experts consider composites the ultimate answer to the utilization of low grade wood. They predict that in 20 years species, quality, and defects of wood will be irrelevant. Instead, wood will function only as the raw material for reconstituted wood products. From a marketing perspective, such predictions are counter-productive. Reconstituted materials can compete only on their merits and not merely as a convenient way to use low grade wood.

The benefits of composites differ from those of solid wood. The benefits combine those of each component in the composite. The new material may perform better, have the advantage of lower weight and volume, may be less expensive to manufacture, be chemically resistant, or less prone to decay.

Any wood product that is glued together or reconstituted from wood fiber is called a wood composite. The wood composite industry has focused mainly on producing low-cost or price-driven panel commodities—plywood, oriented strand board (OSB), medium density fiberboard (MDF), waferboard, insulating board, hardboard, and particleboard. These commodities are processed to value-added secondary products, such as moulding and millwork, and a host of miscellaneous items like clothes hangers and closet organizers.

The manufacturing of composite wood panels is so new that the United States Census Bureau did not create a standard industrial classification (SIC) for this industry until 1987. (The SIC code for these panels is 2493—Reconstituted Wood Products.)

Combining wood with other materials, such as metals, glass, plastics, and synthetic fibers, produces composites that together perform much better than the individual constituents. Some of these wood/nonwood composites are described in Figure II-8.

One composite that is taking hold in the construction industry is the relatively new structural insulated (foam core) panel.

The majority of the rigid foam panels on the market are expanded polystyrene rigid foam cores laminated, or sandwiched, between interior and exterior sheathing material. They may also be urethane foams either glue-laminated to sheathing or foamed in place.

Their energy efficiency and sound absorbing qualities, coupled with the ease and speed of construction of foam panel homes, are winning customers in residential and commercial construction. Superior energy factors have earned structural insulated foam panels a place in projects such as polar research stations, desert installations, basement walls, condominiums, nursing homes and residences all over the world. The success of these panels has led to the formation of the Structural Insulated Panel Association (SIPA) organized to advance the interests of the industry.

Strategies for Marketing Composites

The marketing of innovative composites usually follows one of the following two strategies:

- ❑ Become a price leader by making the product at less cost.
- ❑ Use the better performance of the composite product as a differential benefit of the product.

FIGURE II-8
Composites—New Products from Blending Old Materials

Composites that combine wood with nonwood materials—wood and metals, wood and plastics, wood and glass, or wood and synthetic fibers have desirable properties that cannot be obtained from either material alone.

Wood/metal composites combine toughness with improved temperature resistance and strength. Wood and plastic combine in any of three processes that use the synergy between the materials to yield products much better than either wood or plastic alone. One process combines wood matrix/methacrylate plastic to form composites presently on the market as flooring materials, musical instruments, sporting equipment, and decorative objects.

When wood and plastic are mixed together, the plastic material melts, with each component remaining distinctly separate, producing wood/plastic composites. The resulting products are lighter with improved acoustical, impact, and heat reformability properties. They also have the potential of controlled biodegradability—a critical element in plastic utilization—and can be made at lower cost than comparable plastic products. The potential for utilizing lower-cost recycled resins adds to their competitiveness for use in packaging, furniture, housing, and automobiles. The third process alloys wood with polymers to produce a new homogeneous material. Japan is said to be intensively researching this process.

Wood and glass and wood and synthetic fiber composites also produce unique products. Wood and synthetic fiber composites, for example, are used for interior automotive parts for doors, dash boards, and rear decks.

Combining wood with other materials to form new composites promises outstanding opportunities for introducing new products. Companies planning composites wrestle with the question: "Shall we introduce the new product into a market we know or to a new market?" Professor James Utterback of the Massachusetts Institute of Technology told a 1988 conference at the Forest Products Laboratory that companies do best when they introduce new products into markets with which they are very familiar. He urged using a differentiation strategy to market specialized products (i.e., niche products). Recognizing that many of the innovative products are used by the intermediate or industrial market in making consumer products, Utterback further urged using a differentiation strategy that emphasizes the following.

- ❑ The unique performance of the product in solving the customer's problems, rather than merely replacing or substituting it for competitive products.

- ❑ Tailoring the product to meet the customer's production requirements; asking for and using buyers' suggestions.
- ❑ Producing components that minimize cutting and fitting by the customer.
- ❑ Adopting just-in-time processing and delivery practices to help the customer maintain low inventories.
- ❑ Product quality and distinctive features.

Among the recommendations for marketing new products, Utterback suggested:

- ❑ Developing performance and lifecycle guarantees that help customers market their products. The factory-built home industry (an umbrella term for manufactured homes, mobile homes, modular homes) is developing new ways of financing construction based on guarantees.
- ❑ Developing industrial specifications, standards, and tests for the new products. Manufacturers of structural insulated foam core panels have found that obtaining a model code listing from code agencies such as Council of American Building Officials (CABO), International Conference of Building Officials (ICBO), Building Officials and code Administrators International, Inc. (BOCA), and others is opening doors to sales that would otherwise be closed.
- ❑ Working with new or smaller firms that may be more willing to risk using new products.
- ❑ Holding seminars and workshops to demonstrate handling and utilization of innovative products.

Branding Composites

Branding is one of the tactics used to differentiate a product from its competitors. It establishes a product's position in the market. A brand name cultivates and maintains consumer awareness. Branding is not often practiced in marketing solid wood commodities and plywood, probably because it is difficult to differentiate one piece of lumber of the same grade and size from another. Prices of commodity lumber are set in the manner of auctions, and branding would have little impact on the prices a primary manufacturer receives.

However, many of the new reconstituted products—OSB, waferboard, and oriented waferboard,—are now branded for differentiation. Almost all North American oriented strand board/waferboard manufacturers are employing brand names as a differentiation strategy. In 1988 Weyerhaeuser Company introduced a strand product called Structureframe, specifically for high stress applications in upholstered furniture frames such as front and back rails of sofas. Performance is emphasized in the Structureframe branding.

Branding marks a change in the philosophy of marketing the new family of composites as specialty products rather than commodities.

A "pull" distribution strategy helps encourage industry to utilize innovative products. The National Association of Home Builders annual display house used laminated veneer lumber and Parallam in its "New American Home." As consumers gain an acquaintance with innovative products and begin to ask retailers and builders specifically for these brands, distributors quicken their interest. Making innovative composites available to home craftsmen, especially those who write columns and books for handymen, develops familiarity with properties and increases the comfort level consumers have in using these products. Workshops and seminars for do-it-yourself consumers in home center stores raise interest in new products and reduce fear of the unknown.

Selecting New Products for Development

The Achilles heel of new product development is marketing: management becomes enamored of an idea and spends time and money developing the new product. Almost as an afterthought, consideration is given to how the product will get into the hands of customers, or even whether the customer wants it. The marketing oriented company practices reverse engineering—starting with the customer wants and engineering the product to fill those desires.

Establishing criteria for measuring success is as critical as good design. Bringing out a new product is often a process of persevering despite frustrating obstacles. How wasteful for a firm to introduce a new product only to learn that it fails to meet internal requirements for profit margins or that the competition already has a corner on the market!

Hundreds of products and markets are available. Selecting the successful new product rests on how accurately a company can answer six basic questions.

- ❑ Who are our customers?
- ❑ Which of their needs do we intend to satisfy?
- ❑ How will our customers learn about the product?
- ❑ Are other companies making the same or similar product?
- ❑ How will the product be made available to the customer?
- ❑ How much will that customer be willing to pay for the product?

Answering those questions introduces, in turn, new questions. Figure II-3 (pages 30–31) is a new product selection checklist. The checklist serves to alert the company to potential marketing limitations.

Introducing the New Product

The six basic questions in introducing a new product become six tasks a marketing manager faces:

- ❑ Clarifying how the product meets the needs of the market segment selected
- ❑ Creating customer awareness of the new product
- ❑ Providing information to customers on how to use the product
- ❑ Positioning the product to attract customers
- ❑ Inducing the customer to take a chance on a new product
- ❑ Having the product available when, where, and how the customer wants it

The reasoning of a typical firm illustrates how a company goes about accomplishing these tasks:

> This company currently manufactures moulding and other millwork items. It plans to add a product line of decorative solid wood panels with attached matching wood moulding to be distributed to do-it-yourself consumers through retail lumberyards and home center stores:
>
> ❑ Clarifying how the product meets the needs of the market segment selected
>
> > The targeted customers, women and the over-65 age segment of both sexes share similar needs. They have limited strength abilities, therefore do-it-yourself projects must be within their physical capabilities. They have limited time, for different reasons. Cost may be a factor, but if the product helps them tackle a project they would otherwise have to hire out, product cost may be secondary. Both segments are interested in improving their surroundings. The average woman has difficulty handling panels exceeding 4 feet in length and two feet in width and weighing more than 25 pounds. Few customers in this category have

trucks or vehicles for transporting long lengths. The panels should fit in the trunk of a car.

❑ Creating customer awareness of the new product

The budget includes funds for intense promotion activity to create a demand that will "pull" the product through distribution channels. Some "push" may also be required to induce retailers to feature a new wood panel.

A publicity story on how anyone can be his or her own decorator is effective promotion. Featuring the ease of installation of the product, the woods used and their decorative qualities, the story is slated for the home section supplement published by many newspapers. The published story will be reprinted and used to alert wholesalers and retailers that customers will ask for the product. The company will also cooperate in a promotion program with the home center stores.

❑ Informing customers how to use the product

Clear and complete instructions assume that the customer does not know the basics of carpentry. The panel package includes all hardware for assembly. Any assembly that calls for finger tight screws is tested by persons with fingers with the strength of an average woman, rather than those of a strong young man.

❑ Positioning the product to attract customers

Point-of-purchase tags and display signs are used to help position the product. Attractive packaging and printed information support the quality and value image of the panel packages.

❑ Inducing the customer to take a chance on a new product

Promotion and positioning induce a customer to chance a new quality product more than price discounts or coupons. But retailers and wholesalers may be induced to push the product by getting higher margins. Margins set higher than wholesalers usually realize and additional discounts for retailers compensate in part for the risks and efforts required to sell the new product.

❑ Having the product available when, where, and how the customer wants it

Retailers link manufacturers to customers. A program to assure dependability and quality of the panels gains the confidence of retailers. Panels packaged to assure that the merchandise arrives in good condition and that they easily fit in storage and shelving units encourage the retailer to trust in the manufacturer. (See Chapter VI To Market, To Market . . .)

STRATEGIES FOR THE ESTABLISHED PRODUCT

In the growth stage of a product's lifecycle, sales of established products increase rapidly. Paper industry producers of disposable diapers experienced a phenomenal growth rate of 30 percent a year for ten years before the product settled into maturity. The growth period is the make-or-break stage in which the company establishes the competitive position of the product. This is also the stage in which the product should earn sufficient profit to repay the investment for development.

Strategies to maintain a favorable competitive position in the growth stage include:

- ❑ Establishing the product's differential advantages
- ❑ Developing brand identification
- ❑ Maintaining close relations with retailers
- ❑ Experimenting with the influence of the product price on sales volume. Reducing the price of some products, especially those of prestige value, may serve only to impair the quality image. The sales of other products may exhibit price elasticity, that is, increase with a decrease in price, or decrease with a price increase.
- ❑ Establishing cost leadership by continually lowering prices
- ❑ Investing in advertising to create a strong brand preference

A smaller firm may adopt the same competitive strategy as the better financed industry leader. If the large firms are promoting just-in-time delivery and custom packaging, smaller firms might promote similar features, providing, of course, they can deliver on their promises.

Sooner or later, products settle into maturity. Sales neither increase nor decrease noticeably. The market potential may change, as when the birthrate falls. During a baby boom the sales curve of wooden cribs and baby furniture shoots up. As the birthrate flattens, the sales curve flattens. Or, new competitors enter the baby furniture field. Or a cost leader may slash prices. To hold its share of the market, a firm needs a strong marketing program. Companies have many choices to maintain sales in the mature stage:

- ❑ Proliferate products
- ❑ Segment markets
- ❑ Attempt to change the product image
- ❑ Extend the mature stage

Some companies have been remarkably successful in extending the mature stage of established products.

> The producer of Barbie dolls extended the maturity stage by adding an entire line of complementary dolls and accessories—Barbie's boyfriend, her pets, clothes, beauty aids—to entice little girls into longtime alliances with Barbie dolls.
>
> Food processors play on the public's anxiety about cholesterol and fats in its diet by heralding those foods lower in cholesterol and fat to extend their mature product lines.
>
> AIM toothpaste established a position in cavity prevention with its boast of 50 percent more fluoride.
>
> Other companies have found success in developing a new image for the product. Changing the image of blue jeans from utilitarian to high fashion clothes illustrates: blue jeans, long considered workingman's clothes, were redesigned as fashion pants and took on a new life, reaching an even higher sales plateau.

Finishing, staining, and packaging lumber in individual shrink-wrapped packages to attract new remodelers has been a successful tactic of wood products manufacturers in finding a new segment of customers thereby extending the mature phase.

The industrial product, nylon, used in women's hosiery is a classic illustration of developing new uses for a product. When the hosiery market appeared to mature, nylon was used in broadwoven fabrics; before that tapered off, nylon found its way into the broader market of outerwear and carpets.

III

CUSTOMERS ARE KINGS

. . . who they are
. . . what they want
. . . why they buy

CHAPTER SUMMARY *Marketing a wood product focuses on the customer rather than the product. The customer is king, and the wants and needs of this king influence all the activities of our businesses. Recognizing and understanding the expectations of customers when they buy products is the essence of marketing wood products. This chapter describes who these kings are, what they want, why they buy, and how to use information about customers to increase sales.*

Changes in the population mix, social values, and lifestyles of consumers have splintered the mass market kingdom of yesterday into hundreds of new market segments, fragments, and "particles," each with its own wants and expectations. Business now must create products and services to meet those very diverse expectations. Matching products and services to distinct market groups is the way to the customer's heart. This chapter presents six methods for segmenting the millions of consumers into marketing groups. It also describes ten ways to target industrial markets. Suggestions are offered for developing lists of potential customers and minimizing risks in segmenting markets.

Customers are people. They may be working mothers stopping in the supermarket to buy dinner. They may be working fathers stopping in the lumber yard after picking up the children from day care. They may be double income couples with no kids—the DINKS, older retired persons with lots of leisure time, or youngsters with money to spend. Or they may be the purchasing agent for hardwood lumber at a furniture plant, or in a government agency buying pallets, or a manager of a manufacturing plant. Whoever they are and whatever they do, customers are people, not merely census numbers or entries in a computer program.

In modern marketing, companies no longer think of themselves as merely producers of products. Instead they are working to gain customers who are willing to buy and have the money to buy. These customers influence what a company produces and sells more than technological capability, material resources, or any other factor. The customers are the kings in modern free enterprise business.

Customers are similar to a shuffled deck of playing cards, with the deck representing the 240 million American consumers—give or take a few million. Today, marketing resembles a game of poker. Players work at getting an orderly winning hand out of the shuffled deck. Marketers aim at obtaining a semblance of order from the hands they are dealt by the economic, demographic, ethnic, education, and experience diversity of today's customers. That diversity leads to market segments that share enough similar characteristics to be recognized as distinct buying groups with specific wants and needs. The media like to nickname these groups. Seniors, nuclear families, yuppies, working mothers, baby boomers, working men, single parents, and traditionalist women represent the "in" groups of the moment.

Yesterday's mass market corresponded to an unshuffled deck, arranged in neat order according to suite and card number. Traditional middle class consumers, the mainstay of the mass production market, shared common goals and values, needs and wants. Marketing poker could be played according to comfortable rules, assured of a good chance for a straight flush or a royal flush. Today's consumers have splintered into literally hundreds of markets, each with different sets of needs and wants and expectations. Marketers cannot depend only on Lady Luck to fill a hand from this new deck. The staggering new diversity of today's consumers has changed the way we do marketing. We now divide the mass market into bite size pieces. It's a new poker game out there.

In the 1970s, these divisions were called *market segments.* Market segmentation implied an orderly division into groups. By the 1980s, these segments broke into chaotic market *fragments.* In the 1990s, these fragments are disintegrating further into particle markets, reflecting the growing demographic, economic, and cultural diversity of America.

Whereas marketers in the 1970s could segment women into housewives and working women, in the 1980s they were fragmented by economics, age, and lifestyle into the yuppies, DINKS, and other groups. In the 1990s, these fragments are becoming smaller particles with different spending behaviors. Clearly, as one out of five married couples become step-families whose demands differ from the old nuclear families, and as the baby boomers reach the age of empty nesters, Americans will spend their money differently. Add to these particles, the growing Asian and Hispanic populations with their cultural diversity, and it is apparent that predicting customer attitudes and customer spending behavior has become infinitely more complicated than ever.

Although the difference between segments, fragments, and particles is recognized, they are all called segments in this chapter to simplify discussion.

Market segmentation is an important and fruitful undertaking for any company. It provides ten readily identifiable benefits for reaching customers:

- Segmentation helps a company benefit from change. Markets are dynamic and ever changing. Segmentation tracks the changes.
- It enables a company to cash in on opportunities opened up by new technology.
- Segmenting markets helps meet or beat competition by uncovering the segments with greatest market potential. For example, consumer spending on home remodeling is expected to shift because the number of married couples with children under 18 headed by someone aged 35 to 54 is projected to rise by nearly 3 million between 1986 and the year 2000. During the same period families headed by someone under age 35 is expected to decrease by more than 3 million.
- Promotion, packaging, and distribution dollars are spent more effectively with segmentation.
- Segmentation helps identify additional markets for existing products. For example, the manufacturer of a cloth covered particleboard casket found that the material could also be used by the busy consumer segment interested in attractive, organized closets.
- Segmentation helps confirm or refute hunches about markets. A marketer may sense a new trend to more "gingerbread" in home decorating. Which segment of the remodeling market has adopted that trend?
- Segmenting affords an opportunity to observe factors that contribute to the success of competitors.
- Segmenting the market provides information on the product advantages that help a company's customers survive and compete.
- Gaps in serving certain markets may be discovered in segmenting. This gap can lead to finding a market niche.
- A deliberate process, segmentation helps systematize the process of gaining a competitive edge.

Master Lock Company provides a model of the way a company benefits from segmenting its market. An ordinary lock is a utilitarian item often sold as a commodity. By segmenting the market, Master Lock was able to capture the young customers who buy 40 percent of all bicycle locks. Targeting that segment, Master Lock redesigned the lock with a more stylish look. The Company repackaged the bicycle lock in a flashy red and black design and renamed it "Tough Stripes." To position the locks for younger bicyclists, Master Lock promoted "Tough Stripes" through MTV media and expanded distribution into specialty bike stores frequented by this market segment.

WHO ARE CUSTOMERS?

In modern marketing, the customer is king. Although each customer is an individual, marketing attempts to segment them into groups with similar characteristics and some common needs and wants. From a marketing standpoint, these groups can be expected to buy a product that meets those needs. Defining these groups is *market segmentation* and the segmented market becomes the *target market.*

Market segmentation sometimes is an afterthought prompted by lagging sales. But segmentation ought to be part of the manufacturer's product planning process. The target segment may be restricted to the domestic market, or part of the global export market. And events and politics in Asia, or Europe, or other sectors of the world profoundly impact American forest products marketing. Defining and understanding market segments forces firms to look at more sophisticated methods of marketing than "gut feelings," on which many entrepreneurs rely.

Purchasing Behavior

Three types of customers buy wood products:

- *Consumers* buy wood products for personal use.
- *Industries* buy raw materials such as lumber or components for use in the manufacture of products for resale.
- Construction *contractors* buy lumber and lumber products for use in construction or remodeling and repair.

Customers also include enterprises that use or resell items not consumed in producing other products. These may be brokers or wholesalers, commercial enterprises, or organizations such as governments or institutions. When the federal government purchases wood pallets for shipping or universities purchase wood pellets for heating they are buying products for their own use. A product may also have a dual market, sometimes purchased by consumers and sometimes by industries. A board is an industrial product when purchased green by a furniture manufacturer; it becomes a consumer product when dried, finished, shrink-wrapped and sold to a homeowner who will use it to make a bookcase for personal use.

The buying behavior of industrial customers differs from those of consumers. A hardwood flooring manufacturer purchases lumber when contractors for new housing and remodelers order hardwood flooring. This is a *derived demand,* based on the demands for new housing and repair and remodeling. Some industrial products are related to the demand for other industrial products. The demand for veneer, which is linked to the demand for furniture, is a *joint demand.*

Inventory policies also influence purchasing patterns. Industrial and construction purchases are affected by the trend toward just-in-time (JIT) deliveries. The general state of the industrial economy affects *demand variability.* The demand for industrial wood pallets varies with industrial activity, and is linked to the price of the wood used in manufacturing the pallets. Industrial buying decisions regarding the purchase of *installations* use different criteria than those used in buying *accessory equipment, fabricated parts and materials, raw materials,* and *supplies.*

Consumers' buying habits depend on whether they are purchasing *convenience products, shopping products,* or *specialty products.* As the forest industry moves from selling to marketing industrial products, these classifications often blur. A sawmill may supply both lumber (raw material) and fabricated parts (hardwood and softwood dimension parts). As engineered wood—structural I-beams made of oriented strandboard (OSB) topped and bottomed with solid lumber—takes over a larger share of the structural building products market, installations and accessory equipment may become part of the sale.

The turnkey approach opens an opportunity for smaller firms to use their unique skills to find a niche in the construction and repair and remodel industry. The niche could take the form of painting or laminating lumber, providing custom services, repackaging into smaller than standard loads, manufacturing alternate sizes, and other specialties.

SEGMENTING THE CONSUMER MARKET

Segmenting consumers into smaller groups is a necessity in a market with millions of customers as diverse as the American market. Dividing this huge total market into smaller groups has come to be considered the only realistic way to view the bundle of needs and wants of consumers.

Psychologists and sociologists have attempted to segment markets by classifying the needs of people in some orderly manner. One analysis used in marketing divides basic human needs into the fourteen classifications shown in Figure III-1. Manufacturers of a wood product destined for consumers can hone in on a market segment by discovering how that product matches at least one of these basic needs.

There are five methods most frequently used to sort the millions of consumers into smaller market groups:

- ❑ Demographic differences
- ❑ Geographic differences
- ❑ Psychographic differences
- ❑ Product attributes
- ❑ Product benefits

Demographic differences are based on sex, age, ethnic group, education, income, occupation, family size and status, religion, home ownership, age of children. High income families buy differently from lower income families; Families with children have different interests from young professionals. Figure III-2 shows some of the characteristics that make each group unique.

Geographical differences are based on where people live: urbanites, suburbanites, rural folk, large cities, small cities, inner cities, north, south, east or west, cold climate, mild climate, hot climate. Urbanites living in a city apartment are not likely candidates for garden furniture; a company marketing a product to insure security would look to a retirement community.

Psychographic differences are often the "wild card" in marketing. Psychographics separate consumers into groups that share similar lifestyles, behavioral patterns, activities, interests, and opinions (called *AOI statements*), and value systems.

> Psychographic analysis recognizes that consumers may buy the same product for different reasons.
>
> Consider marketing bicycles: To one set of customers, a bicycle represents merely transportation; to another it is a health product; and a third may consider it sports equipment. The marketer asks "What are customers really buying from me?" Similarly, wood garden furniture may have a different meaning for some people than others. To one set of customers, it may represent leisure. To another, home decoration. To a third, furniture for the children to play on.
>
> The attempt to learn why consumers buy certain products has led to many sophisticated classifications of consumers according to their value systems. The classifications illuminate buying behavior. A few are included in Figure III-3. In 1978, Stanford Research Institute International (SRI) introduced the *VALS System* (values and lifestyles), which classified consumers into nine different behavior types. Finding that the link between values and lifestyles and what consumers actually purchased was not as strong as originally

FIGURE III-1
The Basic Human Needs

The marketer's task is to discover the best match between the product or service provided and one or two of these basic human needs:

❑ Self esteem—People satisfy their need to be noticed by buying unusual clothes and personal items.

❑ Recognition—People want a sense of superiority, which they nourish by buying and displaying items ranging from expensive furniture to professional symbols.

❑ Achievement—From home improvement to courses in success, people purchase help in accomplishing challenging tasks.

❑ Independence—People buy distinctive hair styles, brand name cars, or unique furniture to demonstrate their independence.

❑ Affiliation—People buy items or services that make them feel like they belong in the social or business group of their choice.

❑ Dominance—The need to feel powerful leads to purchases of items ranging from powerful cars to power clothes.

❑ Care Giving—Most people care for someone or something—children, pets, the helpless, the frail aged, or plants. They buy products and services to help meet their care giving needs.

❑ Care Receiving—People also want to receive care and look for personal services and items that meet this basic need.

❑ Stimulation—People buy items with intense sights, sounds, scents flavors and textures to meet the need for stimulation.

❑ Consistency—People need a sense of order and buy items that match or coordinate, or items like closet organizers that help bring sense of order.

❑ Diversion—People buy entertainment to distract them from routine problems and cares.

❑ Security—People want to reduce the risks of harm, danger, injury, or illness and buy products or services that make them feel safer.

❑ Novelty—People buy items that provide a break from the routine and the mundane.

❑ Understanding—People need to understand and interpret the events happening in the world around them and buy news media and instructional materials to meet this need.

FIGURE III-2
Profile of the American Family

1989 Profile	Millions
Total Families	64.8
White	55.6
Black	7.2
Hispanic Origin	4.7
Married—Couple Families	51.0
Wife & Husband Employed	25.2
Husband Only Employed	11.1
Wife Only Employed	2.5
Families Maintained by Women	10.9
Families Maintained by Men	2.8
Families with Kids Under 6	14.5
Married Couples	11.5
Maintained by Women	2.6
Maintained by Men	0.3
Families with Kids 0–17	16.9
Married Couples	12.3
Maintained by Women	3.8
Maintained by Men	0.7

Source: Deseret News June 25-26, 1990

expected, SRI more recently reclassified consumers into eight behavior types, termed VALS 2. The behaviors are measured according to income, education, self-confidence, health, intelligence, eagerness to buy, and energy level. VALS 2 relates a person's values and lifestyles to current buying behavior.

Stanford now offers marketers a service to analyze a cross-section of the product market, determine the psychological type, and then look for media and mailing lists with a high concentration of the VALS 2 types to which the product appeals.

The expense relative to market share deters smaller companies from using formal psychographic segmentation. But a close study of the advertisements and other promotions by large wood products marketers reveals the psychographic segments they have targeted for their wood products. An informal assessment of the psychographics of the target market pays off in further insight into the product benefits to be featured, the best method of distribution for reaching that market, and how to promote the product.

FIGURE III-3

Some Psychographic Classifications of Consumers

Personality types: Six types, including outgoing optimists and conscientious vigilantes

The Homemaker types: Practical women, convenience-oriented women, the economy-minded, the traditional, and the experts

The Technologically Advanced Families (TAFs): Families ready to purchase cable television, home banking, electronic shopping, and other high-technology products and services earlier than other families

THE VALS SYSTEM—CONSUMER VALUES AND LIFESTYLES ACCORDING TO STANFORD RESEARCH INSTITUTE

Strugglers—with a median age of 61 and the lowest incomes of any group, they tend to be brand-loyal consumers.

Achievers—successful, work oriented persons who favor established products and services that show their success to their friends and relatives.

Strivers—whose values are similar to Achievers, but don't have the money to buy what they would like to have. They put faith in style as they try to be like the people they wish they were.

Fulfilleds—mature, responsible, principled, educated professionals, open to new ideas and change. Though they have high incomes, they are practical customers who appreciate quality.

Believers—with more modest incomes than Fulfilleds. Believers are principled, conservative, and predictable consumers who tend to buy American products and established brands.

Experiencers—younger, action oriented, energetic consumers, who spend heavily on new clothing, fast food, music, and other youthful favorites.

Makers—practical self-sufficient people, with little interest in the broader world. As consumers they buy products with a practical or functional purpose.

Actualizers—enjoying the highest incomes, high self-esteem and enough money to buy the finer things of life. They value image as evidence of their good taste, independence, and character.

A smaller firm may reasonably feel that psychographics is a marketing tool only for large companies with marketing budgets. But take a second look at the VALS classification in Figure III-3. A manufacturer producing a low priced home storage cabinet may be weighing promotion to gain brand name recognition. According to VALS 2, however, this low price item might appeal to strugglers and believers. Both of these groups tend to purchase brands they know. Budgeting funds to gain brand name recognition might be a wise investment.

Psychographics lures professors and marketing professionals into devising new systems of classifying people. As the older citizens market takes center stage, experts have come up with a psychographic segmentation of older Americans. The National Association for Senior Living Industries (NASLI) divides the older population into six psychographic segments: martyrs, explorers, pragmatists, adapters, preservers, and attainers. Another classification of mature Americans was devised by Georgia State University, as shown in Figure III-4.

These categorizations tempt marketers into easy generalizations. But businesses are cautioned that the buying characteristics of groups will undoubtedly change as we move into the 21st Century. By 2005, the tidal wave of baby boomers who turn 40 in 1990 will reach the age of 55. As older adults, their purchasing habits will not be inhibited by the same Depression-era concerns as their parents and grandparents. Studies have shown that the current crop of 55 and older adults would rather save than spend. Will the upcoming older segment share that philosophy?

Product attributes—who uses the product—provide another method for determining the segment of the market to target. There is little incentive for reaching non-users of the product, or to promoting to those who

FIGURE III-4

Psychographic Segmentation of Americans Aged 55 and Over

Georgia State University's Center for Mature Consumer Studies classifies the 55 and over population into:

❑ Healthy Hermits—persons in good health, but with little interest in socializing. They are an especially good market for do-it-yourself products and can be best reached through direct mail and print media.

❑ Ailing Outgoers—health-conscious persons in poor health but still socially active—a good segment for leisure products, best reached by selling related products at the same time (cross selling), sales promotions, and special services.

❑ Frail Recluses—persons in poor health and socially and psychologically withdrawn. They are a good market for home entertainment furniture and furnishings that increase comfort and are reached through mass media and selling related products.

❑ Healthy Indulgers—outgoing and in good health, still want to live well and do new things. They are a prosperous market for entertainment and high-tech products, effectively reached through store special promotions, specialized periodicals, and direct mail.

seldom use the product. Segmentation by usage and product benefits helps assure that marketing dollars spent culminate in sales. Segmentation by quality, styling, brand name, warranties, packaging and other product attributes considers the features consumers want. They are characteristics intrinsic to the product itself, usually measurable and observable. Product attributes also include how often customers buy a product and how much they usually buy.

Product benefits—what benefits the buyer expects to obtain from a particular purchase—reveal the value the buyer places on a product.

> Years ago, market research conducted by Daniel Yankelovich revealed that the watch industry at that time had little understanding of buyer's expectations about watches. Relatively expensive watches were being priced and designed to appeal to achievers and prestige seekers and marketed through jewelry stores. But Yankelovich's survey indicated that more than two-thirds of the people who owned watches had purchased on the basis of durability and price. U.S. Time Company used the survey results to segment the market on the basis of price and product quality. Its Timex watches were distributed through stores frequented by buyers who were price and durability conscious—discount houses, drug stores, and variety stores. Within a few years this segmentation approach helped U.S. Time Company become one of the largest watch companies in the world.

Computerized list segmentation is a high-tech method of learning about consumers. Specialized market research firms have devised computer programs that make possible the classification of households in the United States by the 36,000 zip codes and the 206,000 post office carrier routes into a number of homogeneous socioeconomic family groups. Some companies have introduced software packages that tell which individuals on any mailing list are most likely to respond to advertisements. Although these and other computerized public databases currently are used chiefly by large direct-marketing companies, they provide what some marketers call a "no-waste" marketing technique, which should be available to smaller companies in the near future.

Matching a product to the buying behavior of a market segment is a tool for making marketing decisions on:

- ❑ How to package the product
- ❑ How to position the product to attract that segment
- ❑ The differential advantages that will appeal to the segment
- ❑ The most effective promotion package
- ❑ The acceptable price range
- ❑ The distribution channels that the segment uses

FIGURE III-5
Using Market Segmentation to Make Marketing Decisions

The utilitarian table appeals to practical people whose prime interest is functionality—people who are economy minded, probably makers or believers, according to VALS 2.

The expensive table appeals to customers with developed taste, who understand quality and who would like to be considered high-class. They are status conscious persons, achievers, fulfilleds, or actualizers in the VALS 2 classification.

Marketing strategies differ for the two tables:

Item	Packaging	Distribution	Promotion
Utilitarian Table	Ready-to-Assemble (RTA)	Mass Merchandisers	Point of Purchase Women's Magazines on sale in supermarkets
Expensive Table	Assembled; packaged in sturdy boxes	Retail furniture store; interior designers	*Better Homes and Gardens;* upscale home magazines; advertising and publicity

Consider a coffee table. Ready to assemble, with a laminated waterproof top, it can be a utilitarian, relatively inexpensive, piece of occasional furniture. Carefully put together, with a fine oak top with parquet inlays, crafted by artisans, it becomes a prized possession. Who purchases the utilitarian table and who spends more for the expensive coffee table? Figure III-5 shows how market segmentation guides marketing decisions on packaging, distribution, and promotion.

SEGMENTING THE INDUSTRIAL MARKET

Segmenting the industrial market is more easily accomplished than identifying consumer segments. The ten ways most frequently used to segment the industrial market are described in this section.

1. *Geographically—local, regional, domestic, export*

The market for industrial products tends to be more concentrated geographically than the consumer goods markets. Processors of furniture hardwood dimension and components cluster around furniture manufacturers. The availability of the timber resource and aggressive promotion programs recently spurred the redevelopment of furniture manufacturing in several states.

Geographic segmentation is common for bulky products with low value to bulk ratio. A firm that specializes in wood parts with a relatively large value-to-bulk ratio is not constrained by geographical limitations, though proximity to the wood resource is preferable.

2. ***By the number of buyers***

The producer of an industrial product suffers the disadvantage of a limited number of buyers. A producer of hardwood furniture dimension has a finite number of furniture manufacturing firms as its customer base. If the producer markets custom components, the number of potential customers is even more limited. Some products are targeted to larger firms.

An industrial product may be produced for a captive market—wood blocks for a regional toymaker, pallets for a grocery chain, lattice for a garden supply distributor. Captive markets bring with them the advantages of a secure, known market. They also carry the danger of a severely limited market.

3. ***By wood species—softwoods, hardwoods, exotic woods, utility species***

Primary manufacturers segment according to customers who require certain species, certain grades, fiber sources, veneer sources; this segmentation resembles the practice of the textile industry which segments customers into those who use natural fibers such as cotton and wool and those who use synthetics.

4. ***By product type—solid wood, panels, composites, fiber, chips***

Primary producers segment customers on the requirements for form—solid wood or composites, particleboard for underlayment, hardwood plywood for cabinets. Customers also may purchase chips or fiber.

5. ***By size—sales volume, number of employees, plant capacity, and similar size criteria***

A producer of hardwood dimension components may target medium size furniture manufacturers without facilities to process rough mill lumber, rather than larger companies that have their own rough end processing facilities.

6. ***By product attributes—kiln-dried, green, price, quality, durability, construction, degree of processing***

Customers are segmented according to specific moisture requirements. Others demand high quality or shrink-wrapping; still others are unforgiving about deviations from dimensional requirements, and some are segmented by bar codes requirements. Those who buy on price alone form another segment.

7. ***By method of distribution—direct sales, original equipment manufacturers, distribution channels, including retailers, wholesalers, jobber, manufacturers agents***

Weight or size may limit items to be sold through direct mail however, many premium products are successfully sold through direct mail. General Foods, for example, found direct mail a good distribution system for its gourmet foods. Some toys, on the other hand, are sold only through specialty toy stores or department stores. Other products are more appropriately marketed through mass merchandisers. Specialty niche items such as wood tongue depressors are distributed through medical supply channels.

8. ***By industry affiliation—hardwood dimension, furniture, pallets, heavy industry, prefabricated housing, cabinets***

Identification of the industry in which customers are affiliated offers a quick and clear segmentation. A firm making pallet components would segment the market into pallet manufacturers and pallet repair businesses. Manufacturers of dimension components segment furniture and cabinet manufacturers according to price, style, geographical and other specific requirements. Truss manufacturers look to the several segments of the factory-built home manufacturing industry.

9. ***By price***

Some buyers buy inexpensive wood items to use as give-away incentives. In this case, price is the segmenting criteria. Others limit purchases of high priced prestige items for Christmas gifts to clients. Fast food stores want inexpensive give-away gadgets.

10. ***By country***

Segmenting markets by country recognizes the diversity of standards, customs, preferences, regulatory environment, climate, and trade practices. A particleboard manufacturer would hesitate to target exports to countries with strict formaldehyde limitations. The proposed European standard, said to be the most stringent in the world, allows a free formaldehyde level of no more than 10 milligrams/100grams and would be open only to companies producing low-formaldehyde particleboard.

Industrial purchasing is a serious business. Most industrial customers employ a purchasing agent or designate a management person to handle buying. The ultimate decisions may involve several levels of management and more than one department. The economic realities of industrial purchasing put a manufacturer's marketing strategy to test. Figure III-6 illustrates the many levels involved in an industrial purchas-

FIGURE III-6
The Making of an Industrial Purchasing Decision

A manufacturer of redwood garden furniture is approached by a cedar lumber salesman. The cedar salesman has spotted a potential customer. The garden furniture manufacturer faces a possible shortage of redwood. The salesman sees a potential opportunity for substituting cedar.

Production, productivity, engineering and design, marketing, perhaps the maintenance department, as well as the purchasing agent may experience repercussions from the substitution. The plant manager is involved in the decision, as are the marketing manager, the furniture designer, and the president of the company. The sales presentation must satisfy the perspective of each of these decision makers.

The plant manager is concerned with the tactical advantages of the cedar. Will the chipper be able to handle the residue? What are the worker safety implications? What about dust?

The designer wonders if cedar will accommodate the current popularity of the softer, rounded edges in rustic furniture. With sales of resin furniture on the rise, wood garden furniture should adopt a modern design to be competitive.

The marketing manager is anxious about the reaction of customers to the change. The firm's promotion literature now concentrates on the advantages of redwood and the company image is built around redwood.

The president of the company is concerned about the bottom line. How will the change affect profitability? Does it mesh with the company strategy? Will there be a reliable supply of cedar? Will switching to cedar save an investment in new equipment required to manufacture plastic or resin garden furniture?

The purchasing agent wrestles with problems of delivery, quality, reliability of supplier, and price.

ing decision. To capture an order, the salesman must make not one but many presentations. The differential advantages that persuade a plant manager differ from the advantages that appeal to the president or a marketing manager. The "canned" approach won't do.

Finding Potential Customers

Segmentation paints markets with a broad brush. Converting marketing ideas into sales means finding flesh-and-blood potential customers. Larger companies may employ Micromarketing, utilizing mail order lists, census data, and historical buying patterns to assemble a precisely targeted prospect list.

Established businesses might find potential customers in their own office files. One direct, simple method is to analyze sales receipts to develop a profile of buyers for a product.

Marketing an industrial product requires two segmentations: first, identifying the target market; second, identifying the concerns that need to be addressed for each segment of purchasing decisionmakers.

> Joan Silver, president of Reeves Audio Visual Systems, Inc., turned analysis of her company's sales receipts into a targeted marketing plan. When Reeves Audio Visual began business, 90 percent of the customers were *Fortune 500* companies. Sales receipt analysis showed that in 15 years the revenue from these *Fortune 500* customers accounted for just 58 percent of sales. Smaller companies, some with under $1 million in sales, accounted for the rest of the growth. Further analyses revealed a profile of users and pointed to new market segments tailor-made for their products. Assistance in compiling a list of potential industrial customers is available free of charge from most libraries. Specialists also provide lists for a fee. Figure III-7 lists some sources of assistance.

A PROGRAM FOR EXPANDING A CUSTOMER BASE

A real estate broker who rose in three years from salesman to president of a prosperous real estate firm disclosed the secret of his success: "I've found that customers are everywhere and every time," he explained in a TV interview. His ten-point program, shown in Figure III-8, is equally effective as a guide to expanding a customer base in marketing wood products or services. It is high impact marketing at its best.

MINIMIZING RISKS IN MARKET SEGMENTATION

Attractive and promising as market segmentation is, there are risks. Minimize risks by applying a dash of common sense and by drawing on the experience of others. A segmented market may be too small for the

FIGURE III-7
Sources of Assistance in Compiling a List of Potential Customers for an Industrial Product

(The notation "paid service" indicates that the data is available for a fee or price.)

- ❑ Most states publish directories of industries within the state, usually according to the United States Census Standard Industrial Classification (SIC) codes.
- ❑ States with forest industries publish directories of primary and secondary manufacturers, usually available upon request.
- ❑ Bibliographies of information about companies that may be potential customers are available at many libraries.
- ❑ Dun & Bradstreet's "Market Identifiers" provides data on firms by SIC classification, including employees, annual sales volume, and name and title of chief executive (paid service).
- ❑ "Business Information Sources", is published by the University of California Press, 223 Fulton Street, Berkeley, CA 94720.
- ❑ "Directory of Industry Data Sources", is published by Ballinger Publishing Co., Cambridge, MA 02138.
- ❑ "Directory of Directories", is available from Gale Research Co., Book Tower, Detroit, MI 48226.
- ❑ "Thomas Register of American Manufacturers", published by Thomas Publishing Company, New York, N.Y. lists over 100,000 manufacturers by product classification. It is available at most libraries.
- ❑ Dun & Bradstreet's "Million Dollar Directory" is not always readily available at libraries, but is useful for identifying individual companies with more than $500,000 net worth (paid service).
- ❑ Standard & Poor's "Register of Corporations, Directors and Executives," Standard & Poor's Corporation, New York, N.Y. includes details on individual companies (paid service).
- ❑ Forest Industry Directories: Miller Freeman Publications of San Francisco, California; Miller Publications of Memphis, Tennessee; Random Lengths of Eugene, Oregon; and other firms specializing in the wood industry publish directories of producers of primary and secondary products. Some directories cover only hardwoods, some softwoods, some secondary manufacturers.
- ❑ Associations publish directories of their members. For example, The Hardwood Association, National Association of Dimension Manufacturers, and American Plywood Association also publish directories. Addresses of major associations are listed in the annual "Directory of Forest Industries."

FIGURE III-7
Sources of Assistance in Compiling a List of Potential Customers for an Industrial Product (Continued)

❑ Trade journals in the forest industry usually publish annual directories of manufacturers. Some directories are regional. For example, *The Northern Logger and Timber Processor* includes operations in the northeastern states. Others are listed by industry. Some examples are: *Pallet Enterprise,* which lists pallet manufacturers and suppliers of pallet manufacturers. *Furniture Design & Manufacturing* publishes an annual directory edition of both suppliers and manufacturers. *Forest Industries* publishes several issues a year with directories.

company. A small market may be ideal for a niche marketer but a disaster for a firm that relies on what has been named "Chinese marketing." Chinese marketing earned its title from a book published during the depression of the 1930s. Carl Crow, in his book "Four Hundred Million Customers" proposed that the United States could sell itself out of the depression by selling a United States made product to each of the

FIGURE III-8
A Ten-Point Program for Expanding a Customer Base

1. Contact present customers. Are they familiar with the entire product line?
2. Contact past customers. Why haven't they bought recently?
3. Contact contractors, builders, distributors, gift stores. Do they know about the product line?
4. Keep in touch with government, civic, and organizational leaders who may know about similar projects.
5. Use the Yellow Pages to learn who is selling or handling the same type of product or service.
6. List professions or persons who would use a similar product or service.
7. Arrange to speak on a related topic at business clubs, civic clubs, and other organizations looking for speakers.
8. Attend industry meetings and learn about customers for similar and competitive products.
9. Exhibit at trade shows and shows of related products and services.
10. Distribute business cards and literature frequently and widely.

estimated 400 million Chinese. The scheme suggested that if we could just sell one unit to half of the 400 million Chinese we would earn our way into prosperity. "Chinese marketers" look at massive numbers and say wishfully, "If we could just sell to 1 percent of this massive market." It's the same fallacy that might prompt a high-end furniture manufacturer to advertise during a Super Bowl telecast because it has such a huge audience, rather than in *Better Homes and Gardens* with a smaller but sharply targeted audience.

A technology-based company can easily fall into the trap of developing equipment for a small, specialized market.

One high-technology company designed and developed an expensive advanced-state-of-the-art piece of computer equipment and found that the market consisted of only one customer who could use the equipment. The product was too sophisticated, too high tech for the average customer. A competitor came out with a simpler, less complicated device that met customer's needs, and the market segment for that equipment grew as if by magic.

Basing segmentation on stale data may lead to markets that are diminishing or standing still. Misinterpretation of customers needs results in segmentation mistakes.

> When Texas Instruments (TI) introduced a watch for joggers, it added a stopwatch capability. But the watch was designed to recycle from zero every 15 minutes, hardly convenient for a jogger on an hour run. The choice of a jogger market segment was not at fault. TI simply neglected to learn its customers' needs.

IV

MARKETS

. . . finding segments, niches, and trends

CHAPTER SUMMARY Yesterday's single mass market has been replaced by hundreds of special markets, each representing the needs of small segments of the population. Among these segments are groups whose uniqueness and importance to the wood industry merit individual attention. After assessing the reality of "growth markets," this chapter examines the characteristics of special markets—"niche" markets, governments, the specifier market, and global markets. Analysis of the high-priority homeowner and do-it-yourself markets sheds new light on opportunities for the wood-based industry. The craftsman market and the demands of the contractor market likewise open up possibilities for wood products expansion.

A section on how to find new markets and techniques for spotting market trends provides a guide for leaving no stone unturned in the search for the competitive edge.

Hundreds of smaller market segments are replacing the old mass market of middle class America. That market, with millions of persons shared relatively similar values and wants, lent itself to simpler products, distribution, promotion, and pricing. And that market could be reached comfortably with standard marketing techniques. No longer. Standard techniques are applied with difficulty to the dissimilar markets that have replaced the undifferentiated mass market.

Every market—be it the baby market, the child market, the older consumer market, ethnic markets, DINKs (Double-Income, No Kids-couples), or the myriad industrial markets—has unique characteristics that affect the product design, promotion, distribution, and pricing.

Industrial markets, which have always been segmented, have likewise splintered into additional segments, each calling for individualized marketing techniques to match their specialized requirements.

A few of the significant industrial and consumer markets are described in this chapter to demonstrate segmentation and its value in selecting the marketing techniques that produce the best value for marketing dollars.

THE "GROWTH MARKET"—MYTH OR REALITY?

Studies crowd the shelves of businesses, industrial developers, and government policy makers in search of "growth markets." Growth markets that assure an ever-rising sales curve for products are as elusive as the pot of gold at the end of the rainbow. Markets are not static; they can best be described as fragile. They do grow, but they also decline and change. Markets may grow, but the industries and businesses making the products and providing services for these markets only grow as they meet the demands of customers.

> Professor Theodore Levitt, writing in the *Harvard Business Review* (September-October 1975), observed: "There is no such thing as a growth industry. Industries that assume themselves to be riding on some automatic growth escalator invariably descend into stagnation. The history of every dead and dying growth industry shows a self-deceiving cycle of bountiful expansion and undetected decay.
>
> "To continue growing, companies must ascertain and act on their customers' needs and desires, not bank on the presumptive longevity of their products."

A company or an area that believes its fame and fortune lies in a magic growth industry that guarantees continuous and perpetual sales may suffer the fate of the railroads, laundries, and old corner grocery stores.

Railroad use declined not because passenger and freight transportation needs declined, Levitt points out, but because the railroads did not have the vision to fill the need themselves. By defining themselves narrowly as being in the railroad business, instead of the transportation business, management let cars, trucks, and airplanes take away their customers. Levitt says that these industries declined because they were product-oriented, rather than customer-oriented.

A "growth" market is a reality. A "growth" industry may be a myth.

There are myths associated with markets and industries, according to Levitt. These myths are:

- ❑ The belief that growth is assured by an expanding and more affluent population.
- ❑ The belief that there is no competitive substitute for the industry's major product.
- ❑ Too much faith in mass production and in the advantages of rapidly declining unit costs as output rises.

- ❑ Preoccupation with product . . . improvement and manufacturing cost reduction.

Is there a lesson for the forest products industry in those four beliefs? Is it possible that what happened to the railroads could happen to the forest products industry?

Dr. Norman E. Johnson, Vice President of Research and Engineering for Weyerhaeuser Company, told an audience at an international Forest Products Research Society meeting that sawmills producing commodity lumber are not paying attention to customer's needs. From his woodworking hobby experience, Johnson cited difficulty in buying straight boards with structural consistency, free from warp and twist, that won't rot, and that will take surface finishes. Dr. Johnson warned that if we don't do a better job of learning how wood performs in place, and educating the public about how to use wood for best results, other materials will take over. He cited a Rubbermaid Company survey that found outdoor decks the fastest growing product of the decade. Rubbermaid is already marketing Rubbermaid planters. Will outdoor decks of Rubbermaid be next?

Following the reasoning of Levitt and other marketing professionals, recognition that their mission is providing shelter, rather than just lumber, and focusing on customer needs, rather than more and more production of the same product, could save the American forest products industry from the fate of the railroads. The expansive and visionary concept of being in the shelter industry, rather than the lumber business or the furniture business, opens doors that the narrower definition closes.

Understanding the needs and wants of customers unlocks the door to growth markets. Customer segmentation is the path to finding these markets.

NICHE MARKETS

Niche markets offer opportunities for small and large wood-based businesses, industrial and consumer products, and companies that are timber rich or timber poor.

Niche markets are narrow market segments that have become lifesavers in many industries. Manufacturers of any size can gain and hold a share of customers by serving a little corner of a product market, or "niche," as the French say. American furniture manufacturers that survived the debilitating inroads of foreign competition often succeeded by finding a niche in the furniture market. Almost every issue of the chemical journals, agricultural publications, and business magazines relates the success stories of textiles, pharmaceuticals, and other industries in moving from general commodity markets to niche markets.

Every year *Forbes* Magazine lists the 200 best small companies in America. Appropriately called its "Niche List," the annual tally includes companies that have found their niche in products such as superpremium ice cream, custom coded shelf labels, and uncooked pizza. The *Inc* magazine annual selection of its 500 best companies also leans heavily to ward companies that have found their niche markets in products or services.

A niche market is a special market in which a company has the best potential to succeed. The niche marketer makes a product or offers a service that it can do better or more efficiently than anyone else. The French use niche to describe a corner in which something fits. Webster's dictionary defines it as something for which a person is best fitted—a place, employment, or activity. Niches may be as small or as large as a company can manage. Mercedes Benz is a company that markets to a niche. It is also the largest manufacturing corporation in Germany. Niches may be defined in terms of geographical boundaries as well as size: The single European Economic Community scheduled for implementation in 1992 will become a market niche for manufacturers. It will also form the largest market in the world.

Many wood products companies grow from small niches, as did Spectra Furniture Manufacturing Co, State College, Pennsylvania that began with small cherry and walnut items manufactured by a high school student in a basement home workshop. The demand for the items outgrew the home workshop and the student. His parents joined the effort, forming a manufacturing company. In a short time, the business expanded from small pieces to oak and cherry furniture, which found a ready market. Now Spectra Manufacturing occupies a sizeable building and is turning out a complete line of furniture.

Success factors in niche markets may include an available low cost resource or development of special equipment to make a product better or quicker. Or it may fill a price gap in a market, or single out a market with a need that no one else is filling and play it for all its worth.

A successful niche market is one that is:

- ❑ Likely to last. Avoid fad products like hoola hoops or skateboards, unless a distribution system is available to take instant advantage of the fad.
- ❑ Specialized enough so that larger companies are not tempted to swallow them into their product lines. Personal computers were

once a niche product. Now they are part of the products offered by many companies.

- ❑ Clearly defined by product, use, geography, or other parameters, assuring minimal competition. If a product doesn't maintain a dominant leadership position in a market niche, it's time to look for another one.
- ❑ One that a company can win with because it is particularly capable in some aspect of manufacturing or marketing.

Before entering the world of niche markets, a company ought to ask itself these questions:

- ❑ What can we do better than any other company?
- ❑ What can we do faster?
- ❑ What can we do more economically and efficiently than any other company?

With the answers to these questions in hand, the company can then proceed to find its own niche market.

TEN STEPS IN DEVELOPING A NICHE MARKET

Successful niche marketers in the forest products industry have identified ten steps in developing a niche market:

1. ***List possible market niches that might fit the company. Look for:***

- ❑ A local or regional industry that may have special requirements
- ❑ Changes in consumer market segments
- ❑ Changes in lifestyles
- ❑ New regulatory requirements
- ❑ Customers with technical problems
- ❑ Gaps in price categories
- ❑ Markets that are not served or underserved by other manufacturers
- ❑ The special needs of distribution channels
- ❑ Spin-off products

Examples of some of these niches:

- ❑ The market for high end, single family detached manufactured homes, is reported to be underserved.

❑ Swing Designs of Framingham, Massachusetts found its niche in backyard wooden playgrounds for upper income homeowners.

❑ Precoated treated wood deck packages are a natural spin-off for manufacturers of outdoor deck material. Do-it-yourself wood lattice gazebos are a natural spin-off for a lattice manufacturer.

❑ Treated wood shelf systems to withstand the moisture conditions created by plants in the popular solariums. Products that solve problems.

❑ Furnishings and accessories to withstand too much sun in the popular sunrooms. Furnishings that will withstand dark and dampness in basement playrooms.

❑ Improvement of the environment is the foundation of a new class of products in which the niche is environmental acceptability.

❑ Despite the increasing attraction of electronic toys, quality wood toys still enjoy a market. The Brio Scandtoy Corporation, of Milwaukee, Wisconsin serves that market with a line of 197 toys made from hardwoods and skillfully marketed.

2. *Identify raw material.*

The forest products industry differs from computers and other industries in its dependency on a natural resource—wood. Access to timber is essential for certain niches. If the timber is close by, specialized primary wood products may be the niche. If timber is not at hand, the preferable niche may be in secondary wood products. The first step in identifying a niche is to determine the raw material available. Look particularly for an inexpensive source of raw material that may provide a competitive edge in manufacturing costs. Such raw material may come from:

❑ Low value or waste wood

❑ Proximity to a source of panel boards—a nearby plywood plant, particle board plant, OSB, waferboard or other panelboard facility

❑ Waste material from other primary or secondary operations—trim ends, short logs, low grades

❑ Process waste such as dust, chips, sawdust, slabs

❑ Logging waste—thinnings, branches, biomass

❑ Urban waste—discarded pallets, wood refuse

❑ Availability of a preferred species also may provide a competitive cost edge.

❑ Underutilized species

3. *Make a preliminary short list of niches.*

Match the market opportunity in Step 1 with the resource opportunity in Step 2 to make a short list of niches. Learn about their markets, the investment requirements, entry timeliness, reasonable risks, economies of scale.

4. *Identify company strengths and weaknesses.*

Do the niches in the preliminary list match management interest and experience? Specialized equipment, processing capability, a distribution system, access to output that can become the raw material base, skilled labor, and marketing skills may be the success factors for a niche product. Unused plant capacity that can be converted to manufacturing the niche product may yield lower manufacture costs.

Automated Builder, the trade journal for the factory-built homes industry, conducted a survey to discover where the top 100 component manufacturers planned to fill in sales when the construction market slowed. The survey showed a trend to ward finding niche markets in trusses, door components, door frames, and other products ranging from gazebos to garages.

5. *Identify the factors in the niche that meet company objectives, and determine the company willingness to commit to meeting the requirements:*

Market size, growth rate, potential profitability, ease of entry, technology changes, minimum regulatory restrictions, minimum competition, opportunity for export, and similar criteria might be considered in selecting the niche.

Does the company have access to necessary finances, technological competence, and marketing know-how? Are the product and market compatible with the company mission and objectives? Does a responsible executive have the time to devote to developing a niche market?

6. *Narrow the preliminary list of niches by matching market requirements with company ability and willingness to meet these requirements.*

List the preliminary niches on the left and company criteria on the right. If the company is looking for, say, opportunity to export and one of the preliminary niche possibilities is recycling and rebuilding special pallets for which export potential is not readily apparent, remove that niche or move it to a low priority. If the investment requirements in new equipment are higher than the company can afford, cross that niche off the preliminary list.

7. *Learn from other companies the critical areas for success with these niche products.*

Does success depend on service, design, reliability, ease of use, speed of delivery, flexibility, low cost, price, distribution, advance technologies, basic technological excellence? Does the company have the capability to provide these services?

8. *Decide how to strengthen the company's competitive position in critical areas.*

Make this decision after discovering the strengths and weaknesses of competitors. Do competitors perform well in all critical areas? Where is the chink in their armor? A competitor may be weak in distribution and service. Can the company excel in these areas?

9. *Ask for advice.*

Customers, a staff member who has had experience in the market niche, suppliers, distributors, consultants, perhaps another firm with the potential for a joint venture—these are all sources of advice.

10. *Keep alert to the fate of products out in the market.*

In his home workshop, Michigan entrepreneur Kenneth Watts crafted wooden toys with large wheels. The toys were aimed at the children's market, with grandparents who shopped in malls as the likely buyers. The push toys with the large wheels did not capture the grandparent market as expected. However, they did sell—not as toys, but as back massagers! The entrepreneur changed the marketing focus from toys to gifts for men. In the men's department, the new back massagers became a popular item. So successful was the back massager that Watts organized Creative Wonders, Inc. and expanded out of his garage to a plant. He progressed to distribution through a chain of novelty stores, where the back massagers sold well. Except in one store. Not content to let the poor performance stand, he visited the store and found the item displayed in the toy department. Moving the item to the men's gift section immediately improved sales.

FINDING A NICHE

Niches are everywhere. They may result from diversification—making new products to reach new or existing markets, or intensification—

adding breadth and depth to an existing product line to penetrate more deeply into a selected market. Manufacturers of storage sheds exemplify intensification by adding larger, attractive wood buildings to the one-size-fits-all utilitarian storage sheds.

Niches may be driven by wood resources, products, markets. They may emerge from internal forces or external conditions. They may be large or small.

Niches may be found in:

- ❑ An available low value or waste wood resource, or a preferred species
- ❑ Proximity to a plywood plant, particle board plant, OSB, waferboard or other panelboard facility
- ❑ The special requirements of an industry
- ❑ Meeting the changing needs of consumer market segments
- ❑ Changes in lifestyles
- ❑ Changes in regulatory requirements
- ❑ Unusual technological capabilities
- ❑ Solutions to customer problems
- ❑ Gaps in price categories
- ❑ Markets that are not served or are underserved by other manufacturers
- ❑ Accommodating to the special needs of wholesalers, retailers, and other distribution channels

How Manufacturers Find Niches

Here are some examples of how companies have created niche markets for themselves.

Manufacturers can create a product to fill a market niche from an available resource that other manufacturers consider low value or waste material, or from preferred species. For example:

> Parsons Pine Products in Ashland, Oregon has developed a prosperous niche business from the scrap lumber that large mills were throwing away, burning, or chopping into chips. Using these scraps, Parsons Pine has become the nation's largest producer of mousetrap bases, shipping carloads of superior mousetrap bases all over the country. Other niche markets the company has filled produced from the scrap from the region's sawmills are shutter slats and siding for doll houses. Being the best, the indicator of a successful niche marketer, has led Parsons Pine to innovations in mill machinery and to unusual personnel policies that recognize workers as an asset.

Goode Douglas Logging, of Kamloops, B.C. is a timber harvester that has developed niche market opportunities out of logging operations. Goode Douglas produces 600,000 tree and orchard props for the California agricultural market from the waste material in the woods that nobody else wants. The company also cuts up whatever waste wood is available, including plywood peeler cores, and bundles it into firewood for the California market.

An East Coast firm uses a bolter mill to produce upholstered furniture cuttings from short, low grade, hardwood logs. The short pieces sold as chips for $60 per cord; as cuttings the sales value was $201 per thousand board feet of cuttings.

A plant in the Lake States uses 3-foot blocks of aspen and birch to produce thousands of tongue depressors, popsicle sticks, and similar items.

Some manufacturers are forced to create niche products when they lose their commodity market. For example:

Some areas have prohibited cedar shingle roofs as fire hazards, leaving cedar manufacturers with a diminished market. Cedar manufacturers have devised niche products to utilize cedar shorts in gardens for edgings, planters, and raised beds.

On the other hand, some companies with access to preferred species can use them to create niche products.

The Mr. Spindle line of decking components uses cedar and redwood to manufacture premium spindles, newel posts, and porch posts for outdoor decks. Promoting "easy installation while maintaining the durability cedar and redwood are famous for," the Mr. Spindle line is packaged in a system that provides a point-of-purchase display.

A company many locate near a primary processing facility to develop its niche.

Sunwood Japan Co. Ltd.'s consideration of a plant to make cabinets for stereo speakers in the rural area of LaGrande, Oregon was based on taking advantage of the proximity to a Boise Cascade particle board plant. The particle board is the major component of the speakers.

Filling special requirements of industry creates niche products. When a large company discovers a reliable, quality supplier, the payoffs are significant for both buyer and seller.

Oscar Bergstrom of International Falls, Minnesota found a niche for a pallet operation in the requirements of a nearby large paper mill for a pallet with special features.

Some specialty wood products companies make niche products for industries, functioning as job shops to produce parts or products to manufacturer's specific requirements.

John Knoell and Sons of New Britain, Pennsylvania, a family owned corporation founded over a 100 years ago, specializes in contract manufacturing. Knoell's products range from Franklin mint coin boxes, through antique telephone cases for a telephone company, to decorative oak lids. The firm originally produced wood commodes. When the advent of indoor plumbing ruined the market for this product, they switched to ice boxes. When refrigeration decimated the market for ice boxes, Knoell switched to television stands. Quality, flexibility, and sensitivity to markets has kept this company in business through many economic and technological changes.

Demographics, geographical mobility, and changing lifestyles create an unending stream of niche markets for consumer and industrial products.

California Closets found its niche in fixtures that help busy persons organize clothing and possessions stored in closets. The market niche was fostered by the large number of women entering the workforce. With more spending power but less disposable time, women with both full time jobs and homemaking responsibilities eagerly welcomed help in organizing their closets to simplify housekeeping.

Manufacturers of storage sheds are now responding to consumer needs for more storage space by expanding their lines to include larger, attractive wood buildings. These larger buildings fill a previously underserved market niche—the need for relatively inexpensive week-end retreats, backyard workshops, or swimming pool dressing rooms. Handy Home Products—a major player in the storage shed industry—has broadened its shed line to include gazebo and spa enclosures in six sizes, playhouses in two sizes, and three sizes of dog houses. They have also developed kits to fill the market niche for attractive, easy-to-build wood buildings. The Handy Home Products Weekender Lodge, 8 feet wide, 16 feet long, and 12 feet high with two built-in lofts, a front porch, three working windows, and a domed skylight, comes in a kit that includes complete instructions down to the amount of paint or stain needed.

Wickes Lumber, provides more than 35 models of wood storage sheds, which it promotes with a ten-page "Wood Storage Sheds" manual.

The H Window Company in Minnesota, a Norwegian company, found its original niche in windows to meet the needs of the Norwegian market, which has more than a third of its land above the Arctic Circle. Expanding into the United States, the company found a niche in serving the lifestyle that prefers integrating indoors with outdoors. They met the need for cleaning large windows by building-in a feature that permits easy cleaning. Sales were further fostered by high insulation qualities.

The advent of the computer and accompanying growth in home offices has led to niche markets in home office furniture. Available in ready-to-assemble (RTA), unpainted, and finished furniture, home office furniture has demonstrated consistent growth.

> Changes in lifestyles are reflected in the renewed popularity of traditional interiors and traditional furniture. Manufacturers of grandfather clocks have found a new niche as a home decoration accessory. First popular when the song, "My Grandfather's Clock" became a hit in 1875, grandfather clocks are again in demand.

Changes in government regulations create market driven niches. Proliferating regulatory requirements for pollution control in painting has led some firms to discontinue painting operations, opening up niche markets for companies who have painting facilities.

> Painted particleboard is desired for some products. This need provided a niche for a company located close to a source of particleboard. The company began specializing in painting particleboard and later added cutting and finishing the painted particleboard into shelving. A painted particleboard shelf sold for about equivalent to $800 per thousand square feet. As a commodity, the particleboard sold for only $205 per thousand square feet.

> Limitations on formaldehyde emissions has created a need for panels that meet these limits. There is also a demand for panels used in making cedar closets. Geo C. Brown Company has found a niche in meeting these two demands with its Cedarstran™ cedar OSB. Sanded smooth to prevent snagg clothes, the 4-foot-by-8-foot panels are made from cedar heartwood without the use of formaldehyde.

Manufacturers with laminating capabilities, or experience with more efficient adhesive systems, or the ability to utilize new microthin veneer products find market niches by providing these unusual technological capabilities to other manufacturers in their areas.

> Goode Douglas formed a wood treating company to manufacture pressure treated fence posts and stakes. Their niche market is custom treating wood for larger sawmills and plywood plants in the area that do not have their own treating facilities.

> Microwood Company, Shippensburg, Pennsylvania produces a superthin veneer using a peeling technology originated in Germany and further developed in the United States. Hardwood or softwood logs and unusual hardwood species are turned into real wood veneers with multiple applications.

> Solariums and sunrooms often develop problems in the summer from exposure to the sun. The need for inexpensive sun shades has created a niche market for lattice manufacturers who provide lattice for that purpose. To block out sun rays, Solar Stat of Miami, Florida has found a market niche with self clinging vinyl glass window tinting that blocks off 88 percent of the ultraviolet rays and 55 percent of the solar energy. Easily applied by laying it over the glass, it is also easily removed and can be stored for the next year.

A growing worldwide problem is waste disposal. The United States has developed a new sensitivity to crowded landfills and hazardous waste. Internationally, pollution and trash problems are magnified by limited land space.

> Wood Exchange Company in Corvallis, Oregon has found a market niche in using wood pallet waste as a source of energy for one of its plants and as a source of parts for rebuilding.
>
> National Gyp-Chipper, a small company with just seven plant workers, found a market niche in solving one of the biggest problems of contractors—getting rid of the junk and waste that clutters construction sites. They developed a machine called The Gyp-Chipper, which pulverizes scrap gypsum board and turns it into an easily transportable powder with potential recycling value. Selling for around $10,000, the Gyp-Chipper takes up the room of about two trash barrels.
>
> Gyp-Chippers niche market approach as a model for turning scrap into usable chips or fiber is nothing new. Gyp-Chipper's contribution is finding and serving a niche of approximately 90,000 companies in the contractor market. This small company has developed skills in servicing these customers that it is now applying to the international market.

Changes in demographics produce new niche market opportunities almost daily. As America's population ages, the need for wheelchair access increases. A portable wood ramp or a do-it-yourself ramp fills that market driven niche. Manufacturers of modular and panel factory-built homes have opportunities in a new market niche-homes with features that make it safer and more convenient for use by elderly or handicapped persons. Design changes such as enlarging doors, increasing lighting, removing door stops, walk-in showers, soft bathtubs, and similar adaptive measures fill a niche by providing solutions to customer problems. Other products that fill this market niche are:

- ❑ Doors easily opened by arthritic fingers
- ❑ Cabinets that minimize bending and reaching
- ❑ Raised garden containers for porches
- ❑ Conveniently placed water faucets

The same problems are also shared by the handicapped of every age. Growing recognition of the problems of the handicapped is creating a market niche. Federal mandates for handicapped access are solidifying that market niche.

(Additional suggestions can be found in The American Association of Retired Persons (AARP) publication "The Do-Able Renewable Home: Making Your Home Fit Your Needs." The publication is available free from AARP Fulfillment D12470, 1909 K St. N.W. Washington, D.C. 20049.)

Gaps in price categories and markets that are not served or are underserved by other manufacturers also provide niche market opportunities.

> One of the reasons unfinished paneling gained sales in a period when finished paneling sales remained relatively flat, was the price gap between finished panels and unfinished plank paneling made from comparable wood. The unfinished paneling, ready for the do-it-yourself installer to finish, has found a niche in filling the price point between the high-end finished panels and the lower priced plank paneling. The unfinished paneling has the advantage over its prefinished and plank paneling competitors. It can be custom stained or painted to fit the homeowner's decor, used in whole sheets, or cut into thirds for wainscotting. They are considered easier to install, and better meet the skill levels of traditional panel buyers.

> American Hockey Stick Co. of Iron Mountain, Michigan, now the nation's second largest hockey stick producer, began in a garage after its founder, Bob Berquist, lost his job as a result of the decline of the steel industry in that area. Now expanded into a 6,000 square foot facility, American Hockey turns out more than 25,000 hockey sticks a year.

Accommodating to special needs of wholesalers, retailers, and other distribution channels creates niche markets. Retailers prefer to sell products they stock, but space is limited inside stores. Displaying complete models of decks, storage buildings, carts, and other bulky items encourages sales. Manufacturers can find market niches in packaging products so they are convenient for the customer to haul, convenient for the store to stock, and promote sales at the same time.

> Lambert Corp. found a niche for its dump carts by repackaging them for easy hauling by customers, and to allow retailers to display and stock the items. Orginially packaged in 14-cubic-foot plain brown cartons, the dump carts were disassembled and repackaged in 4-cubic-foot carton with multicolored graphics displaying consumer information. The new cartons were self-selling, in-store displays. The repackaging lowered transportation and storage costs, provided better information for customers, and enabled the retailer to display 59 cartons in the same space that before held only ten cartons. The repackaging increased sales by 30 percent, according to the company.

SPECIAL MARKETS

Among the hundreds of markets for products made by the forest product industry, some segments have characteristics that distinguish them from other markets. The special markets examined in this chapter merit individual attention for their importance to the wood industry.

Government Markets

Governments at all levels constitute a large market for wood products. About one-third of the nation's gross national product is for government expenditures. Clearly, no industry can afford to ignore such a large market. Although the products may be similar to industrial products, government purchasing regulations set this market apart from others. Many businesses are reluctant to bid on government contracts. The paperwork, bureaucracy, regulations, delays, policy changes, and frequent turnover of procurement personnel cause marketing departments of private business to distance themselves from sales to the government.

A positive side of the government market is the detailed specifications used by government purchasers which help assure that all bidders start on a level playing field. The United States Department of Defense, for example, procures according to Military Standards (MIL-STDS), which specify product, packaging, testing, and other attributes. Every company, large and small, is required to meet the same standards. The federal government buys most off-the-shelf items through the General Services Administration, but there are almost 500 other federal offices that procure materials and products. The Office of Federal Procurement Policy has attempted to minimize red tape in purchasing through the Federal Acquisition Regulation (FAR). Most federal offices have Small Business Set-Asides restricting certain bids to small businesses.

State and local governments develop their own procurement procedures, though procurements funded through federal grants usually follow federal specifications. In recent years, state and local expenditures have increased as a proportion of government procurement. Government purchasing ebbs and flows with changes in the political environment, the economy, and government objectives, but it is always a market worth investigating. For assistance in learning which departments purchase wood products, contact the nearest office of the United States Department of Commerce or the Small Business Administration.

The Specifier Market

A barrier between the manufacturer or distributor and the customer may be erected by the person or organization that specifies materials to be used. The specifier may designate wood species, grades, finish, moisture content and other attributes. The specifications may be drawn by the architect of a project or by engineers. These material specifiers often influence purchasing decisions more than the ultimate customer. With the customer virtually removed from purchasing decisions, reaching the specifying architect or engineer is essential.

Yellow poplar offers an example of the architect's influence in species utilization. Yellow poplar is widely used for architectural mouldings and other millwork.

A 1989 survey of architectural millwork manufacturers revealed that architects and designers often specify pine, when yellow poplar would do as well or better. Although poplar is not as well-suited for a clear finish, according to some millwork manufacturers, as a painted product it has great potential. Millwork producers also suggested that they would use poplar if architects and designers specified it. They urged that architects improve their willingness to accept or specify yellow poplar for paint grade millwork.

The way to accomplish this is for architects to be shown the virtues of poplar. Many architects are unfamiliar with poplar and specify pine for painted millwork, when the less expensive poplar could do the job. In calculating a job, some millwork suppliers base their bids on competitive poplar prices, despite specifications calling for pine, risking their profits under the assumption that architects can later be persuaded to allow the substitution of poplar.

Local and state officials form another barrier in decisions on wood product purchases. A survey on specification practices by State Highway Departments on the utilization of bark in highway landscaping showed highway departments in the southeastern states specifying Douglas fir bark from the west, or by brand name (such as Weyerhaeuser), or other specifications indicating a lack of familiarity with bark and its uses.

The education of purchasing decision makers is one of the missions of species-oriented trade associations. Specifier education should be part of the promotion package of any wood product.

Global Markets

An growing number of wood-based businesses are depending on export to provide new markets for their product lines. Too many American companies have in the past treated export customers as a fall-back market when the domestic market declines. That practice does not work well anymore. Many firms have learned that their competitive edge is dulled by a perspective that remains local or national.

In the forest products industry, as in other basic industries, companies are expanding by manufacturing and marketing on a global scale. Internationally focused firms have for a long time achieved global scale by locating manufacturing facilities in other countries to gain access to low cost labor and the local log and lumber raw material. Now comes the realization that global scale in marketing as well as manufacturing is necessary to compete worldwide. As new telecommunications technology, improved transportation systems, and the disappearance of literal and figurative "walls" make the world smaller, marketing gets larger in scope.

Experienced global marketers suggest that global marketing requires two mentalities: a megamentality—looking at the big global picture—and an eye-on-the-sparrow mentality—paying attention to the needs of individuals. Exporting is further complicated by real differences in needs and wants of the people of other nations—needs and wants that are frequently unlike those of the domestic customers familiar to manufacturers.

The broadened view of global marketing puts a new face on export. To expand in exporting, it's not enough to produce products in metric dimensions. The realization is growing that exporting is but one part of the global marketing scene. As in marketing to domestic customers, technology is only part of the answer; successful exporters find they need to learn about and address individual customer's needs.

Exporting Strategies for Smaller Companies

Larger firms with export departments and the financial resources to manufacture and market globally acquire a competitive edge in the international market. That competitive edge is also available to smaller companies.

One method used by smaller companies is joining forces with an international company to capitalize on their people, contacts, offices, and equipment already in place. The two organizations work together on a project-by-project basis, utilizing the expertise of both. Another method is to set up a system of dealerships and use the services of international marketing experts.

> National Gyp-Chipper of Pflugerville, Texas, with sales of well under $1 million, has prospered by exporting. National Gyp-Chipper initially concentrated on local and domestic sales, but visitors from other countries convinced the company that there was an international market for their chipper.
>
> The company generated export demand by exhibiting at trade shows, placing publicity stories, and advertising in trade journals. Setting up a system of dealerships and engaging an international marketing firm solved many

> of the problems of doing business in unfamiliar territory. These dealers became working "partners," providing advice on product protection through patents, copyrights, trademarks, and cash down payments. Exclusive territories were protected by contracts requiring minimum sales volumes.

The United States Departments of Commerce and Agriculture, state economic development departments, international banks, and university international trade institutes are waiting in the wings to assist wood products manufacturers to expand into international markets. Export freight forwarders provide invaluable information for the asking on Letters of credit, shipping, and customs. Most libraries stock guides to exporting. Use of these services helps smaller companies avoid potential disasters in exporting.

> Mater Engineering received a $200,000 order and letter of credit for a project from an African country. Before accepting the order, the letter of credit was submitted for verification to the international department of the bank. The bank's routine check revealed that it was a worthless forgery and that the agent placing the order had a reputation for chicanery. The bank's advice was free, and the savings substantial.

Membership in one of the forest industry trade associations brings access to export opportunities that a smaller business might miss. An example is the American preoccupation with environmental regulations that sometimes obscures the impact of the environmental movement in other countries.

The boycott of tropical hardwoods in the United Kingdom opens doors for American producers of temperate hardwoods. A 1990 poll showed that 61 percent of United Kingdom consumers declared that they would not purchase furniture made from timber originating in tropical forests. Influential specifiers in the United Kingdom are stipulating that shippers must certify that tropical hardwoods originate from countries with long-term sustainable forestry policies. The American Hardwood Export Council (AHEC) has expanded its campaign to assure United Kingdom customers that United States hardwoods are a sustainable, managed resource.

Trade associations and trade journals carry articles on trends and factors beyond the control of individual businesses. They report regularly on two of the factors that affect exports of wood products: (1) the value of the dollar with respect to foreign currency and (2) tariff negotiations. The future of American wood industry trade relations with Japan hangs on international tariff negotiations. Japanese fire, building, and inspection standards have inhibited sales of American finished wood products. Estimates that the potential annual sale of more than $1 billion in wood products hinges on the success of negotiations in The General Agreement on Tariffs and Trade (GATT) has encouraged vigorous action on the part of the United States.

Two segments of the global market generate exceptional opportunities for the forest industry in the 1990s: Japan and other Pacific Rim countries and the European Economic Community of 1992.

Japan

Economic superpower, Japan, is well on its way to becoming a major importer as well as a powerful exporter. *Gaiatsu,* or external pressure exerted by the United States, is behind the breakthrough in selling to Japan.

Japan has become the world's fastest growing market for imported cars. Luxury houses and pricey leisure facilities are making inroads in the buying habits of some segments of Japanese society. Many observers conclude that the Japanese are finally beginning to live like the wealthy people they are. The ideas of younger Japanese about how to live differ from their parents. A poll conducted by the Tokyo research organization Chuo Chosa Sha showed that only 14 percent of those surveyed believe that the Japanese standard of living is better than that of Americans and Europeans. An impressive 76 percent said they would not be able to buy a house with their present income, but only 35 percent thought the Japanese economy would get better in the future.

> Adolf Hertrich, President of Vanport Mfg Co. Inc., Boring, Oregon provides a model for exporting to the Japanese market. He learned the preferences, needs, and requirements of Japanese customers, whose industries use the metric-sized, quality lumber Vanport ships to Japan. At the mill site, Hertrich built a traditional Japanese Tea House in which he welcomes customers. He solves customer's problems with innovative ideas and meticulous attention to quality.
>
> Vanport Manufacturing has followed a prescription for entering the Japanese market that other successful companies have adopted:

- ❑ Win friends in Japan
- ❑ Provide service and support
- ❑ Invite Japanese customers to visit your plant
- ❑ Hire the Japanese whenever possible
- ❑ Deliver only a superior quality product that meets rigorous quality specifications

In other Pacific Rim countries, rising income is reported to be creating new consumers eager to enjoy the amenities available to industrialized nations. The needs and wants of more than 1.7 billion people on the Asian side of the Pacific Rim (including over a billion Chinese) are changing as people move off farms. Families have fewer children, and have more money to spend on those children they do have. A remarkable change in the status of women almost everywhere in Asia is creating demands undreamed of only a decade ago. Women are remaining in school longer, marrying older, bearing fewer children and having them later, and entering the work force in greater numbers.

Observers of the Asian scene say this change will eventually have an impact similar to that of American women in the workforce.

Developments in the Association of Southeast Asian Nations—Thailand, Malaysia, Singapore, Indonesia, and the Philippines (called ASEAN)—and the newly industrialized countries of Korea, Taiwan, Hong Kong, and Singapore (the NICS for short)—will surely impact United States manufacturers. Pacific Rim watchers are hard pressed to keep current on the rapid transformations of Asia's economies.

It appears that Asian nations are also shedding their dependence on European and North American markets. The growth of an intra-Asian trade bloc is at the same time a threat and a promise for the global competitiveness of United States companies. The region is the fastest growing market in the world. The Asian economies that used to depend on exports to the United States market now export greater percentages to each other. The percentage of intra-Asian trade is expected to reach 55 percent by the year 2000.

A metaphor to describe Asia's new economic realities has been described by *Forbes* magazine (November 26, 1990) as "a V-shaped formation of flying geese in which Japan is the lead goose; followed by Singapore, Korea, Taiwan, and Hong Kong, . . . the ASEAN countries, with China bringing up the rear." *Forbes* concluded: "That there are no American geese in this image should disturb anyone interested in the long-term prospects for U.S. business in the global economy."

For wood products manufacturers, the short-term effects may be diminished shipments of furniture, plywood, and wood products from these countries to the United States. The long-term effects may be a diminished market for the export of wood products to the Pacific Rim.

Europe

The European Economic Community of 1992 (EEC 1992) is another emerging market. Marketing managers are agog with anticipation of more than 300 million customers that can be reached in one common market. The leaders of the 12 individual European national markets have been working diligently to harmonize specifications and deregulate border differences. But will we have but one country to deal with in place of the 12 countries? Jean Quatrezooz, president of a federation of 15 market research companies called INRA (Europe), warns that: "Too many Americans are looking at Europe as homogeneous." He believes that the 1992 Europe will be even more fragmented than it is today. Removing all economic barriers will bring a return of old regional borders and in place of 12 countries, we will have to contend with 1,000 regions.

Not everyone agrees that the single European market will never come into being, but most concur that it will take many lifetimes and that we will no longer be able to look at European consumers as we did before 1992. The per capita spending in three categories of potential wood product consumers—housing, furniture, and other household goods—for the 12 EEC countries varies significantly, as do cultural differences and trade practices. There is no one type of European consumer, according to knowledgeable sources. Europe is a collection of different cultures and habits that differ within each country as well as between countries.

But most marketers believe that the benefits of the EEC market integration should far outweigh the marketing costs. Marketing veterans suggest several short cuts to entering the European market:

- ❏ Do as much background research ahead of time as possible.
- ❏ Visit the region and hire a good interpreter.
- ❏ Segment the market by reason for buying a product, which is more important than how much they spend on the purchase.
- ❏ Hire local employees, preferably those who went to the good schools and who are more likely to know "the right people."
- ❏ Recognize that obtaining market information in Europe is much more difficult than in the United States.
- ❏ European countries take the census at different times and have different rules. If the trade barriers between European countries fall, insistence on the typical letter of credit may prove a competitive disadvantage. Many observers believe that the Eastern European picture will evolve and clarify with time and offer outstanding opportunities for American manufacturers and exporters willing to cope with new bureaucracies and unfamiliar work habits.

Conquering Barriers in the Export and International Trade Market

Businesses are apprehensive about the complexities of the export market—unfamiliar distribution channels, pricing extras, unfamiliar business terms, shipping documents, letters of credit, laws and regulations, and different methods of financing are a new set of hurdles. Although exporting and international trade have been hailed as the panacea for marketing forest products, many firms are reluctant to enter the export fray. Perhaps even more intimidating than the technicalities of handling the business end of export is the difference in the cultural climate between countries.

A good interpreter or a common language can breach the language barrier. Variations in cultural norms, or generally accepted behavior are a more difficult wall to breach. Some differences are shown in Figure IV-1.

These cultural differences carry a strong message: Segmentation of the international market goes much further than the geographic, demographic, psychographic and product categories. Of equal consequence is segmentation by cultural differences, which vary from country to country, within countries and regions, and within large geographic divisions.

An understanding of the political structure of the nations targeted for export plays as much a role in successful marketing as understanding a potential customer's needs. Marketing on an international scale is often an exercise in political influence. Learning who is influential in international trade negotiations and trade politics in both the target market and the United States government is essential. As noted in the section on Megamarketing in Chapter I, politics and power frequently join the four traditional marketing *"P's"* of product, place, promotion, and price in international marketing.

The magnitude of these differences accounts in part for the eagerness of American businesses to enter the global age by forming partnerships with firms in the countries with which they plan to do business.

Few smaller firms have the resources to penetrate the maze of international trade structures. Several sources of assistance are listed in Figure IV-2.

The Homeowner Market

Of the three basics of human life—food, clothing, and shelter—shelter needs are the most demanding and least flexible. A thrifty homemaker may pare a food bill and make do with last year's clothing. But once a homeowning or renting commitment is made, other needs are secondary.

FIGURE IV-1
Generally Accepted American Behavior Often Differs from the Accepted Behavior in Other Nations

Personal values vary: Americans are active; businessmen in some countries have more passive behavior.

Americans are industrious; some other nationals are more leisurely.

Americans are often considered materialistic by more spiritualistic peoples.

What is polite social behavior in one country is disquieting to persons of another. The frequently casual behavior and dress of some Americans is offensive to countries used to more formal attire. The competitiveness that marks American enterprise seems out of place in the more cooperative environment of some other national business communities.

Where Americans value performance, others place greater stock in position. In America, a businessman might not deem it worthy of mention that he has written a book or lectured at the University; in many countries intellectual prowess is more highly prized.

Americans are by and large orderly; many other peoples display random behavior (witness the automobile traffic anarchy in some countries).

The American is by nature optimistic; in many other countries fatalism is a national characteristic.

Progress is treasured by American business; other national groups cherish security.

Americans prefer space between people; in some cultures personal body contact is the norm.

Americans are in a hurry to get down to business; in some countries a business executive wants to be sure the person he is dealing with is a friend before he welcomes him as a supplier.

The homeowner market is sharply segmented by age and income. An understanding of the segmentation reveals the importance of this market to the forest industry. According to the 1987-88 Bureau of Labor Statistics Combined Consumer Expenditure Survey, shelter costs the most for 35 to 44 year olds, three quarters of whom own their homes and spend an average of $5,869 a year. For those with annual incomes under $20,000, well over half are renters and their shelter spending falls considerably under that average. But in the $30,000 a year bracket, home ownership and home costs begin to meet the average for that age group. And a whopping 90 percent of those whose income exceeds $40,000 own

FIGURE IV-2
Sources of assistance in international trade

❑ Trade Associations—Join the trade association that promotes your industry's interests. Most trade associations have acquired or are acquiring an international presence. The Western Wood Products Association (WWPA) has become an effective negotiator in opening markets for United States wood products. The National Forest Products Association (NFPA) is engaged in accessing the international political and trade structure.

❑ State Agencies—Most states have initiated programs to assist their industries in competing in the international market. Assistance ranges from providing information to opening offices in countries selected as customer targets. For example, the Urban Institute completed a study called "Targeting Export Assistance Programs in the Southwest" to examine the abilities of industries in Arkansas, Kansas, Missouri, Oklahoma, and Texas to compete in the international market. The state of Oregon, for example, commissioned the preparation of a "Foreign Terminology and Trade Handbook" to assist forest industry businesses in exporting wood products.

❑ Consultants—Private consultants may be able to provide assistance on trade practices, markets, marketing, technical, engineering and environmental and other concerns. Lists of consultants are found in trade journals and from professional associations.

❑ Universities—Many universities have created international trade institutes to conduct studies and advise industries on penetrating the international market.

(Continued)

their homes and spend an average of $8,960 a year on them. That's about 22 percent of a $40,000 income.

The big sleeper for the 1990s in consumer spending will be the 45 to 54 year olds, whose numbers are expected to increase by 50 percent. More of them will have the highest incomes, and despite enjoying more affluence than any other group, will spend less on shelter as their mortgages mature. This should leave more income for other needs.

The number of affluent households decreases sharply in the 55-64 age group, but more of these are homeowners than in any other age group and they spend an average of 81 percent of their incomes on these homes. This falls to 66 percent after they reach 65, as mortgages are paid off.

American Demographics Magazine (February 1991) calls housing costs the "800 pound gorilla of consumer spending." An examination of the

FIGURE IV-2

Sources of assistance in international trade (Continued)

- ❑ The United States Government—The Department of Commerce and the Department of Agriculture (USDA) provide information and direct advice and assistance. The Foreign Agricultural Service of the USDA receives reports on wood products from embassies and commercial attaches in all parts of the world.
- ❑ The Food and Agriculture Organization (FAO) of the United Nations (UN) publishes reports and studies on forestry and forest industry management all over the world. UNASYLVA, their monthly publication on forestry and forest industries, covers few of the everyday concerns of marketing forest products, but it is an excellent source of general information.
- ❑ Prices, trends, and the intricacies of the markets for forest products are reported in trade publications such as the *Import/Export Wood Purchasing News*, including information on domestic regional and international inventories, prices, collections, sales, and prospects for market expansion.
- ❑ Employees, consultants, and acquaintances who have lived or traveled extensively in the countries targeted for export.
- ❑ Trade publications covering the global forest industry—for example *World Wood* provide information on the forest industry all over the world. National publications such as the *New Zealand Forest Industries* reveal problems, plans, and activities in the forest industry for each country.

shelter burden reveals opportunities for manufacturers of items for homes, home remodeling and repairs, furnishings, and other wood-based items.

The Do-It-Yourself Market

The do-it-yourselfer has evolved into the pacemaker of the remodeling and repair market. Smart merchandising has honed to a fine edge the marketing of wood products to persons who do it themselves or who buy it themselves and hire contractors to do the work. Handy men have been around for a long time. Home center stores, discount stores, mass merchandisers, and other customer focused retailers have expanded the do-it-yourself market to include women, grandmas and grandpas, young couples and single mothers coping with growing children. They have spurred the marketing of products to make repair and remodeling of the home easier and more convenient.

Retailers have led the way in the development of a host of new products. The trade journal *Home Improvement Center* features articles every month about stores that increase sales by innovatively meeting customers' needs and wants. How many of those innovations represent lost opportunities for manufacturers who could be earning more from their products by meeting those needs in the design and production phases?

Home center stores have progressed from merely packaging precut lumber and parts. Some stores now have deck design programs to customize the packages. Monticello Software has introduced a computerized software program that customizes decks and runs off a detailed bill of material while the customer waits. The customer returns home with the design and materials ready to start the deck project. Edward Hines Lumber Co. of Itasca, Illinois now provides these deck design capabilities in all 25 locations.

The do-it-yourself market segment knows what it wants, and is willing to pay for products of value. *Home Improvement Center* frequently conducts surveys of home center consumers. Figure IV-3 is the message consumers sent to manufacturers in response to the question: What can manufacturers do to improve the benefits they provide?

The Home Craftsman Market

Craftsmen who work with wood form a subsegment of the larger do-it-yourself group. Skilled woodworkers who handcraft wood furniture use wood as a vehicle for artistic expression. They work in oak, black cherry, mahogany, maple, ash, and teak. A few do their own logging, cutting, and kiln drying to obtain wood with the precise qualities they want. The craftsmen custom design furniture, cabinets, boats, toys, turnings, and musical instruments. Wood accessories, such as lamps, bowls, boxes, and sculpture, are crafted to feature the wood grain or burl pattern. To this labor intensive market, the cost of the wood is relatively minor as long as the craftsmen find wood with special qualities.

The flourishing arts and crafts movement appeals to persons with a desire for individualism and self expression. In a survey of furniture style preferences, the young homemakers with an interest in "cocooning"—using their homes as a focus for living and entertaining—and with some disposable income to spend on nurturing this interest, expressed a preference for antique furniture. Craftsmen fill that need. Crafted furniture styles range from Oriental to Shaker, Mission, early American, Country, and other creations. Regional shows and exhibitions provide the opportunity to display these artistic wares.

Craftsman produce wood Christmas tree ornaments, key holders, children's toys, towel holders, and other adornments for the home.

FIGURE IV-3
Product Benefits Consumers Want

When asked what manufacturers can do to improve the benefits they provide, a 1990 panel represeting 1,000 households with an average age of 44,and average income of $42,000, sent *Home Improvement Center* a strong message:

- ❑ Write clear instructions
- ❑ Provide an 800 number
- ❑ Make easy-to-use products
- ❑ Make quality products
- ❑ Supply free pamphlets
- ❑ Make sure all parts are included
- ❑ Provide how-to video tapes
- ❑ Have product demonstrations
- ❑ Use interchangeable parts
- ❑ Package in smaller quantities
- ❑ Code products by skill level
- ❑ List required tools for installation

Other than requesting easy-to-use products and quality products, consumers aren't asking for more technology, they are asking for better marketing. The customers have spoken. Is anybody listening?

Products started in basement workshops often grow into sizable businesses. Clearly craftsmen are a market niche for wood parts and pieces.

The Contractor Market

Construction and remodeling contractors form a principal market segment of the wood products industry. As major customers of wood products, contractors' needs and wants have a far reaching impact on the wood building products industry. The majority of homeowners and remodelers are forced to rely on the knowledge and judgment of contractors. The contractor's knowledge or lack of knowledge may make or break a wood product. A contractor may specify aluminum siding in place of wood siding, or aluminum windows rather than wood windows, or one brand of door versus another. The recommendation may result from a happy or unhappy experience with a product, lack of knowledge about using the product, or merely an attempt to save money.

A poorly informed contractor can damage the image of a forest product and can create a major marketing problem. If the contractor fails to use the proper nails in constructing a treated wood deck and the deck begins to fall apart in a few years, the customer is likely to swear never to build a wood deck again. If the contractor uses exposed wood beams that have not been correctly preserved, the reputation of wood is tarnished.

Installation is the weakest link of the wood building material industry. Informing contractors about the correct installation of wood products is a wise investment for any wood products firm. Contractors spend thousands of dollars at a time, and are repeat customers. Forest industry firms with a marketing focus make a contractor's business needs their business. When a contractor like Jendell Construction of Rockville, Maryland reports that his company buys many of their building materials at home center stores because lumber yards carry only basic lumber items, but a home center store carries specialty items and finished products, and has reasonable prices, he is sending a message to the forest industry that good, fast service, along with competitive prices, attract contractor business.

Contractors purchase wood products from three sources:

- ❑ Large home builders may purchase directly from the producer. Some, through backwards integration, may own their sawmills.
- ❑ Lumber yards and lumber dealers
- ❑ Home improvement stores

Manufacturers and distributors may take a leaf from programs developed by lumber yards and home center stores to promote sales to contractors.

Contractors are perennially pressed for time. In Elgin, Illinois, Siegle's Home & Building Centers have come up with a program called LevelExpress to save contractors time. Professional contractors joining the program enjoy services that help expedite their work. They can dial direct to a Siegle staffer trained to understand and assist contractors. Contractors in the program are invited to use Siegle's telephones, fax, and copy machines without charge. Computerized project estimating and job accounting are available to LevelExpress members, along with special credit terms. Siegle's company officers are on call to assist contractors at any time. A phone call is all that's necessary for a contractor to arrange to pick up items like kitchen cabinets from the firm's warehouse.

Major home centers are also creating "contractor nights" when suppliers set up booths or displays to demonstrate their products. Special pricing and promotional materials are given to attending contractors.

These programs create opportunities for tie-ins, cooperative information, and promotion programs by wood products manufacturers. Contractor and consumer education about the use of wood may be the most far reaching step the forest industry undertakes to protect its market share in construction.

Remodelers report that homeowners are willing to pay for name brand building products that are known for high quality. Remodelers forecast that manufacturers will be expected to stand behind their product's performance and replace any products that fail. Homeowners expect contractors to guarantee labor and materials. Contractors look to retailers to guarantee those materials. And retailers look to the manufacturer's guarantee. Contractors also often prefer to order by phone or fax. Trained salespeople who understand the products they sell are the most valued service a retailer or producer provides, according to many contractors.

The Greenville, South Carolina-based Builder Marts of America (BMA) provides management services and buys wood products for 850 independent lumber dealers in 47 states. In this role, the association exerts a prime influence on the building products industry, forming the link between primary and secondary forest products manufacturers and contractors. BMA views the contractor as the most important part of the sales equation. Most BMA dealers orient their marketing to these contractors, who select 80 percent to 90 percent of the building products materials used on a job. Although 75 percent of their dealer sales goes to new home construction, including everything from millwork to prehung doors, BMA has identified the repair and remodeling sector as a $200 million market over the next ten years.

Paul Dodge, president of Forest Products and Millwork for BMA, has devised a program to dispel the disadvantages of trading lumber, at the same time maintaining dealer inventory, minimizing dealer investment, and providing more services to contractors. The BMA buying program influences the producer and provides more value for the consumer. The essence of the program is shown in Figure IV-4.

FICKLE CONSUMERS ALTER MARKETS

The characteristics and buying behavior of markets of special interest to the wood products industry have been described in this chapter. But, in truth, before the ink is dry on a published description of the

FIGURE IV-4
The BMA Dealer Program

Paul Dodge of BMA developed a model for analyzing forest products costs. The model has five components:

- The cost of the wood (a quality determinant)
- The gross profit margin
- Freight costs
- Market opportunity and inventory carrying costs
- Purchasing

The program assists the dealer in controlling the essential elements of price, quality, and service.

By converting BMA dealer purchasing from the single transaction of buying forest products on the open market to programmed buying, BMA can offer dealers lower freight costs and eliminate or minimize the market opportunity and inventory costs. This reduction of the component costs allows the dealer to better control costs and price products more competitively.

Dealers in programmed buying arrangements book a volume of business for the year. When the dealer needs a product, he calls up and orders the volume required. This keeps the investment in inventory to an absolute minimum, reduces freight costs by booking on volume, and frees the dealer to be on the floor selling and managing rather than purchasing.

The program benefits accrue back to the forest products processor. To build market share, the supplier locks up a certain volume of business which he can count on from the participating dealers. This commitment can be at market prices, or at a fixed price.

Most of the 850 independent BMA dealers in 47 states are contractor and remodeler oriented. Contractors select 80 percent to 90 percent of the material used on the job, which ranges from millwork to pre-hung doors.

BMA has adopted four strategies to help dealers position themselves to gain market share in that growing market:

- Separate, stand-alone stores
- Installed projects
- Preferred contractor sales
- Setting aside part of the store as a selection center

The BMA program demonstrates that innovation is possible and necessary in every one of the facets of marketing.

buying behavior of a market segment, that behavior begins to change. Consumer tastes and buying behavior are fickle. As they change, they alter the markets. An activist society and a media hungry for news are accelerating the rate of change.

To keep abreast of the changes, the marketer becomes a trend watcher, frequently holding up a finger to the wind. Marketing success rests on discovering people's attitudes and riding with them, rather than attempting to change human behavior. Although it is a popular belief that marketing—specifically advertising—can change people's behavior, behavior modification is not a marketing skill.

Economists, pundits, news reporters, and intellectuals of every stripe attempt to sound the alert on new trends. The forecasts may not always be correct, but marketers listen. Some examples:

- "America's New Abstinence," is how an issue of *Fortune* magazine summed up the late 1980s trend toward light drinking, light smoking, low calorie and low cholesterol food.

  "Led by maturing baby boomers and fitness fiends," states the article, "Americans are breakfasting on less bacon and fewer eggs, forgoing the lunchtime scotch, and seeing a lot more sparkling water and chicken breasts at dinnertime. The new abstinence has touched the lives of so many people that it is creating havoc in some very large industries."

  The new abstinence is a trend that has forced some growing markets into early maturity. It has changed pricing, promotion tactics, product development, encouraged market segmentation, and initiated niche development.

- America's concern with preserving the forests prompted Wing Industries, Inc., Dallas, Texas, manufacturers of Wing doors and Woodlite LaDoor, to create a "Buy a Door/Plant a Tree" program in which Wing promises to plant a tree in the Yellowstone-area forest for every Wing door consumers purchase.

- To receive an "America's Corporate Conscience Award" from the Council on Economic Priorities is a marketing coup. Award winners have developed exemplary programs for preserving the environment, charitable giving, community action, equal opportunity, and employer responsiveness. The award is a product benefit in the eyes of a growing environmentally aware segment of consumers.

- Awareness of rapidly filling landfills and the thrust for recycling were the impetus for creating new products from recycled material.

Anchor Plastics Ltd. developed Re-Source Lumber from recycled dairy containers that are ground, cleaned and extruded as boards, including 1-by-6 deck planks in 8-foot to 16-foot lengths. The boards come in standard wood colors of cedar and weathered redwood. The recycled product is reported "to have none of the problems of using wood."

Publications are available to help marketers keep up with environmental trends that affect markets and marketing. Typical of the many new "green" publications are:

The Green Consumer Letter-1526 Connecticut Avenue N.W., Washington, D.C. 20036

The Conscious Consumer-700 N. Milwaukee Ave., No. 105, Vernon Hills, Illinois 60061

- ❑ The dramatic change in food and eating habits is only one trend making a difference in marketing. Social changes, such as the rise in the number of never-married people affects buying patterns. The middle class consumer of 15 years ago defies identification today. The face of America and many other countries is changing, altering the way we do business.

How do marketing strategies cope with so many fundamental changes? This news from the *Wall Street Journal,* December 9, 1987, tells this story:

"The Going Gets Tough and Madison Avenue Dumps the Yuppies. Conspicuous Consumption is Passe; Finding New Ways to Move the Merchandise."

In the late 1970s and the 1980s yuppies, young urban upwardly mobile progressive professionals, who wore Rolex President watches and Brooks Brothers suspenders or dressed in power suits and carried expensive Coach Metropolitan bags brought joy to manufacturers of luxury items. By the end of 1987, the *Wall Street Journal* was declaring ". . . yuppies have lost their luster." Madison Avenue advertising gurus replaced yuppies with "couch potatoes," a more traditional segment interested in basic values rather than consumption. This is good news for the manufacturers of furniture and cabinets and home decoration products. A story in the *Los Angeles Times,* December 1988, read:

"Redefining Women; *Good Housekeeping* ad campaign sparks wave of criticism." (The ad theme was based on the assumption that working women were returning to home and family and traditional values.)

And as a final example of how affluent yuppies are growing older, the *Wall Street Journal* responded by declaring that "Marketers Err by Treating Elderly as a Uniform Group. . . . Older people are the nation's richest consumer group and have many needs and wants that are not being met by a business community long geared to the younger segment." This change spells opportunity for manufacturers of all types of products.

These stories illustrate how markets ebb and flow. Manufacturers and distributors who fail to keep abreast of the changing needs and wants miss opportunities for offering innovative products and "packages of benefits" that customers buy.

Manufacturers of wood products may feel immune from drastic swings in trends. Food, after all, has always been subjected to fads. But the rapidity with which attitude and behavior changes are occurring leaves no room for complacency.

The hot breath of public concern with preservation of the forests, waste disposal, global warming, and other environmental issues affects a product as commonplace as disposable baby diapers. The savior of mothers since their introduction in 1961, disposable diapers came to be viewed as despoilers of the environment in the 1980s. Producers were put on notice that these diapers were no longer acceptable and that the old fashioned, environmentally benign cloth diapers were coming back into vogue. As the consumer demand for disposable diapers declines, the demand by diaper manufacturers for the absorbent fluff pulps used in the manufacture of disposable diapers falls. After years of surging growth, sales of disposable diapers are threatened. Disposable diaper manufacturers have answered the threat by increasing research and development to make disposable diapers biodegradable.

As public opinion underlines the environmental movement's proposals to preserve timber, the public perceives the timber supply as a precious resource to be husbanded. Consequently, the pulp and paper industry is being prodded into using more recycled paper. "The largest remaining forest is waste paper," according to an interview with Harry Merlo, President of Louisiana-Pacific, reported in the July 29, 1990 *Oregonian*. Merlo is further quoted as stating, "It has absolutely no environmental problems—spotted owl or otherwise—and you don't even need a chainsaw." L-P FiberBond uses recycled paper and the minerals gypsum and perlite. The company also manufactures two forms of housing insulation from recycled newsprint.

DEVELOPING A "TREND SENSE"

The business leader who spots a trend before everyone else gains a competitive edge. The company with a smart trend detective on its staff has an early opportunity to modify and expand products and services in line with emerging buying patterns.

New trends seem to appear on the scene every day. The essence of being a trend detective is avoiding the false alarms and sensing the real changes in consumer behavior.

New trends can be spotted by market research, using focus groups and computer modeling techniques, but studying the media is the simplest and least expensive way to find clues to emerging trends.

That feel for current and emerging consumer demands, that nose for news that endows a marketer with a sixth sense for coming trends, can be developed by "nosing around" and scanning consumer and trade literature.

> The alert marketer gleans information about trends from the morning newspaper or even scanning magazines in reception rooms while waiting for an appointment.

Five Ways to Separate Trends from Fads

Fads like hula hoops come and go. Trends such as the public's growing environmental awareness have a long-term influence on consumer buying behavior. Media coverage has been shown to change the public's thinking and habits in only a couple of weeks.

There are seven ways to use the news media to spot a significant behavior trend:

- ❑ Follow the National Media Index maintained by the Conference on Issues and Media. The index measures the coverage of news stories in 11 major broadcasting and print news media outlets. For example, the index showed increased coverage of environmental news between 1984 and 1988.
- ❑ Read the shelter magazines and other specialized journals. These publications more directly influence buyers of the products than general media.
- ❑ Start and maintain an index of media coverage of trends important to the company. How often is a trend mentioned? The more often people read about the trend, the more likely they will change their buying behavior accordingly.
- ❑ Read and study trade journals related to products. The trade journals are in the business of spotting trends and forecasting buying behavior. Trend awareness in marketing industrial wood products to industrial customers increases the competitive edge of manufacturers.
- ❑ Detect "trend streams" by following news stories from their origin in technical journals, through popular magazines that aim at consumers

interested in topics like crafts, health, safety, and the environment, and finally to mass market magazines such as *Reader's Digest*. By the time they reach the mass media, they are usually a bona fide trend.

- An early alert on an emerging trend is the appearance of new scientific and technical journals. Typical is the announcement of a new journal, *Global Environmental Change*, started in 1990. The magazine was initiated as "a unique international journal that addresses the human ecological and public policy dimensions of the environmental processes that are threatening the sustainability of life on Earth." This is a certain herald of the globalization of the environmental movement, just as similar American journals confirmed the depth of the environmental movement in the early 1970s.
- Numerous trend newsletters are available that provide forecasts on specific topics. An example among the many newsletters that forecast trends that may change the way people live and work is the *John Naisbitt's Trend Letter*, available on a subscription basis.

The idea of trend watching is new to both manufacturers and the merchandisers who distribute their products. Trend merchandising identifies new items that mass retailers should stock. In 1988, the Target store chain first employed a trend spotter to travel around the world from corporate headquarters in Minneapolis looking for new trends, new fashions, new ideas. A line of mauve and blue carpentry tools for women resulted from detecting the trend of women joining the do-it-yourself movement. Wood products manufacturers can benefit from integrating backwards from marketing to manufacturing to take advantage of the same merchandising trends. The manufacturers who develop the products to meet demands for trends have a ready market.

Product bar codes, automatic scanning, and computers are converting subjective trend watching to a fact-based science. These new technologies are shifting the power to position products from the manufacturer to the retailer. The big makers of household products—leaders like Procter & Gamble and Kraft foods—used to run national promotions that determined which products retailers featured. Now, computerized data pooling tells retailers how most products are selling nationally, regionally, and locally.

Trend detecting programs will find application in the wood products industry, especially in marketing to home centers. Bar coding lumber and wood products was the first step. As the industry moves from a commodities mentality to a marketing mode, computer programs will be adopted to provide trend information on wood products. Some of the media sources that marketers will find useful for spotting new trends are found in Figure IV-5.

FIGURE IV-5

Sources for Developing a "Trend Sense"

- ❑ Trendsetting consumer magazines directed to homeowners

Advertisements that show what other firms in the industry feature; articles that tell what's "in" and "out."

Good Housekeeping *Practical Homeowner*

Better Homes and Gardens

- ❑ Daily newspapers

Advertisements for home furnishings

Decorating supplements

Advertising inserts from home center and building products stores

Syndicated news stories

- ❑ TV marketing programs
- ❑ Trade journals in the wood industry

Example: A reader's letter in *Practical Homeowner* of January 1991 asks: "Does any manufacturer of steel insulated entry doors . . . make a replacement attic hatch door? . . . most homes have a 3/4-inch piece of plywood as an attic hatch. If these door manufacturers made an R-15 or better attic hatch *with magnetic weather stripping, they could sell thousands!* " (emphasis by authors)

- ❑ Business-oriented newspapers and periodicals

Wall Street Journal *Fortune*

Business Week *Forbes*

- ❑ Mail order catalogs that include specialty wood products
- ❑ Newsletters

Boardroom Report, major bank economic reports, Federal Reserve Bank bulletins

- ❑ Visiting

Trade shows of wood-based industries—boats, furniture, office equipment, housing

Home center stores, lumber yards, craft stores

V

STANDING OUT FROM THE CROWD

. . . product differentiation and positioning

CHAPTER SUMMARY *The marketing oriented manufacturer designs products to fill customer needs and expectations. A product that uniquely meets customers' needs has a* ***differential advantage****—characteristics that help it stand out from other products. Companies with limited access to capital and limited manufacturing facilities can gain a competitive edge by differentiating products. A product differentiation strategy is equally effective in marketing industrial and consumer wood products. Using the package-of-benefits concept, differentiation can also be applied to indistinguishable commodities products. By their nature, niche products have built-in differentials that influence distribution, promotion, and pricing. Seventeen specific ways to differentiate industrial wood products, and ten methods for differentiating consumer products without the expense of altering manufacturing processes, are described. Additional differentiation techniques are applicable to niche products.*

The explosion of environmental awareness has jeopardized customer preference for wood products. Product differentiation principles are applicable when the public has developed a negative perception of the product.

Positioning is a marketing strategy for creating customer preference for a product. Positioning focuses on the differential advantages for each segment of the market. Manufacturers can choose from seven basic ways to position a product. For example, packaging, brand names, product use, quality, and promotion play important roles in positioning. Positioning can be achieved by providing services. Attention is given to positioning against competition and positioning products for the special segments of the children's market and the mature market.

In the 1970s, cost conscious and environmentally aware consumers returned to the practice of buying food in bulk popular in the days of the old fashioned grocery store. Grocery coops and supermarkets accommodated by making beans and nuts, flour and oatmeal, sugar and salt, and other basic staples available in large containers. Customers could scoop out the desired quantity and pay by the pound rather than the package.

The trend toward bulk buying opened up an unexpected opportunity for the Bradbury Barrel Co., a Bridgewater, Maine producer of northern white cedar barrels. A major manufacturer of utility barrels for storing and handling Maine potatoes, Bradbury Barrel found their barrel sales falling as the potato industry shifted to automated bulk handling equipment in the early 1970s. To utilize its knowledge and facilities for barrel making, the company turned out a new line of barrel furniture, including tables, chairs, cradles, and chests. The barrel furniture line failed in the furniture market and Bradbury continued to look for new markets.

When it exhibited its line of smaller barrel derivatives—rocking horses, magazine racks, cradles, footstools, and similar items at a trade show it stumbled on the bulk buying trend. Incidental to the exhibit was a "tub hi-rise," a half barrel mounted on a stand. Retailers recognized that these barrels were the answer to their need for displaying foods in bulk. Bradbury Barrel quickly responded by positioning the barrels as bulk food containers and as units for point-of-purchase (POP) displays.

When a good product and a need converge, there are always imitators. Competitors for the display barrels emerged. Bradury Barrels needed to find a way to differentiate itself from the competition. The product enjoyed strong ***differential advantages:***

- ❑ Construction of strong, light, rot resistant northern white cedar
- ❑ Brown ash and elm hoops that matched the trend toward "naturals"
- ❑ Tongue and groove construction that allowed flexibility in meeting customer requirements. The barrels could be cut vertically or horizontally.
- ❑ Tubs available in a wide range of sizes
- ❑ Tubs could be finished in a natural look or stained and lacquered for a sophisticated style
- ❑ Plastic liners and covers permitted handling of a wide variety of products
- ❑ The unique history and character of the wood barrel, a singularization interest.

Bradbury Barrel capitalized on the flexibility, diversity, and adaptability of its product by creating "The Idea Barrel." Promotion of the barrels for point of purchase displays used the slogan: "Sell by the barrel . . . . we make buying fun!"

The company offered another differential advantage, one that is essential in marketing a specialty product: It customized barrels to meet customer demand. Imagination became the only limit in using barrels, as illustrated in Figure V-1.

The story of Bradbury Barrel Co. illustrates how product differentiation, featuring recognizable benefits to customers, sets a product apart from the competition.

FIGURE V-1
Bradbury Barrel Uses a Product Differentiation

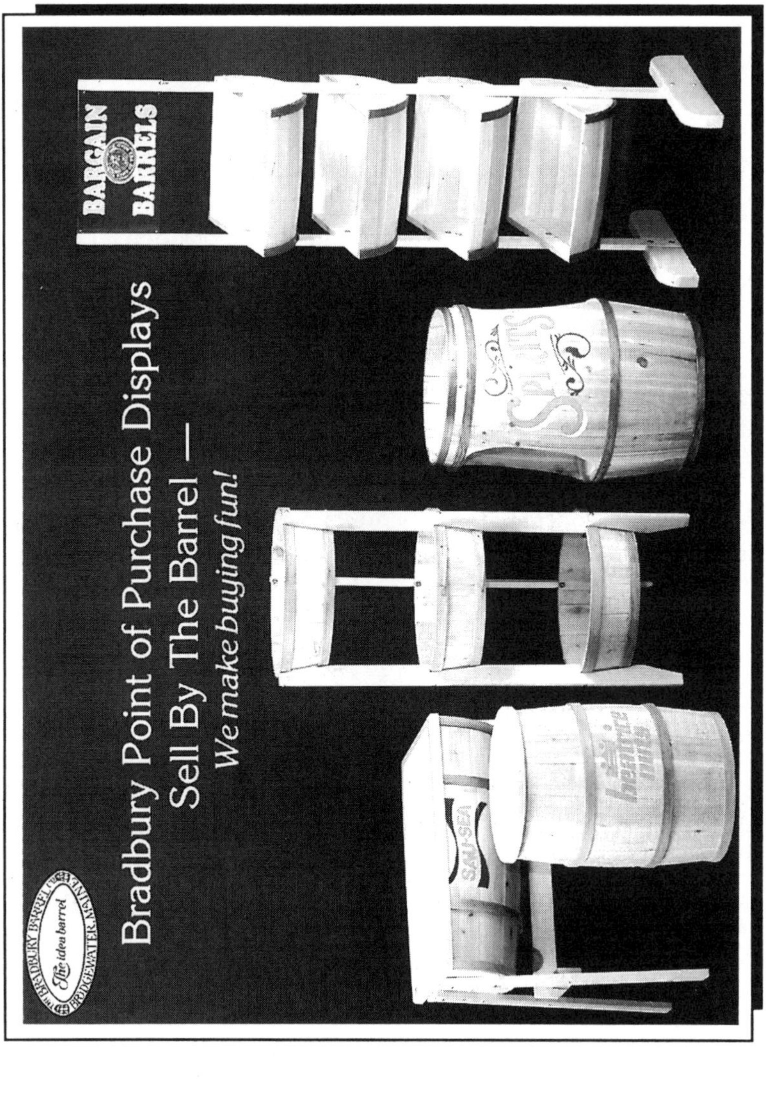

Manufacturers using the differentiation strategy deliberately create differential advantages that buyers will recognize and prefer. Product differentiation is a valuable tool to sharpen the competitive edge of a product. Differentiation helps build brand loyalty and diminishes the importance of price. It also helps insulate a product from competitive rivalry. When customers believe that one product gives them more value than others, or greater status or prestige than others, price becomes a secondary consideration.

Differentiation focuses on the product's unique features, providing a measure of protection against competition. Differentiation is the supply side answer to the demands created by customer's needs and wants. It is a strategy that guides decisions on product characteristics, packaging, brand name, and other product attributes that position a product for targeted market segments.

Promotion is the means of informing customers about the differential advantages of a product.

TECHNIQUES FOR DIFFERENTIATING PRODUCTS

Why does a customer buy one product in preference to another? Because the selected product offers a real or perceived advantage—unique features available only from the product the customer selects. "Uniqueness" itself is a product differential.

A product differentiation strategy is equally useful for industrial and consumer products, but the tactics differ. The concept of product as "a package of benefits" simplifies the selection of product differentials for industrial (including construction) products.

Differentiating Industrial Products

Industrial product differentiation may be achieved in small increments, with significant value added that increases income from the product without expensive changes in processing. A construction commodity such as a 2-inch by 4-inch stud may be converted into a higher value identifiable product, earning additional income for the mill as a specialty product.

> Triple T Studs of Cascadia, Oregon converted its commodity stud production to differentiated specialty products by custom cutting the length of the studs to customers' orders. By manufacturing precision sawn studs usable by the contractor with no further work at the job site, Triple T tapped into a higher value construction market. The company followed the marketing strategy of differentiating a product for a specific market segment. Triple T ties its technology to its markets, engineering in reverse from the market demands to production practices. The only change made in the mill to

accommodate the market demand was to move the trimmer from the end of the sawmill line to the front of the planer.

Another sawmill converted its commodity stud production to a specialty product by painting both ends of the studs, grading out #1 KD (kiln dried) studs, bar coding, packing for the retail market, quoting delivered prices to retailers, and providing mixed loads. Quality and reliability were the differentials featured in their promotion program.

Price as a Differential Advantage

Price is a differential advantage in marketing industrial or construction products. To successfully compete on price, the selling price must drive every manufacturing and marketing decision the company makes on quality, service, and performance. Everything the company does has to be consistent with the chosen price level.

The conventional wisdom today asserts that a company can't compete only on price anymore. Requirements for service, just-in-time (JIT) deliveries, and marketing assistance drive up the price. What appears to be a pricing dilemma is often not about price, it is about the value of the product to the customer. Sawmills sell lumber by the carload to brokers at one price per thousand board feet. Put that lumber into small, easy to handle packages with just enough lumber to build storage shelves and the consumer is usually willing to pay a considerably higher price for the convenience and availability. The consumer benefits because the price of the kit is still less expensive than a finished shelf. The manufacturer benefits because value has been added to the raw material.

Non-price Product Differentials

Industrial wood products are differentiated as readily as consumer products. A supplier of industrial wood products may select from a menu of non-price differential advantages, each tied to a benefit for the purchaser. Sometimes manufacturers are not even aware of the advantages that prompt purchases of their products. A study by market researcher Eric Von Hippel showed that at least 25 percent of the innovative ideas for new industrial products were suggested to manufacturers by users of the product.

Seventeen ways to differentiate industrial wood products with the goal of increasing sales are listed in this section. Each technique has merits:

1. ***Adding new or innovative features not available from competitive products***

Providing the purchaser clearly sees the value of the benefits, the technology to make an unusual product is a valuable differential advantage. Oregon Lumber Co.'s unique "fumed" oak process, which makes new wood look old has opened markets for this Portland, Oregon-based industrial wood products firm. Old buildings are being restored all around the country. Many were originally built with rich looking fumed oak flooring, a product that had not been made for 25 years. To fill a contract for fumed oak flooring for an expensive restoration project, Oregon Lumber had to reinvent the fumed oak process. Now, the company fills the niche for fumed oak by providing the differential benefit of exceptional quality wood flooring for restored mansions and for new buildings.

2. ***Showing how the product helps save the customer production time or costs***

Differential advantages can be developed by understanding how a product is used and showing how purchasing the product helps save the customer's production time or costs. For example, accounting records might demonstrate how the product reduces the purchaser's labor costs. Accounting records that demonstrate that *use of the product reduces the purchaser's labor costs,* i.e. furniture dimension components reduce the furniture manufacturer's labor cost.

A differential advantage can also be created when the product reduces a purchaser's investment costs. The National Dimension Manufacturers Association, for example, retained a major accounting firm to analyze this differential feature. The study showed that purchasing dimension components decreased the investment requirements of a furniture or cabinet manufacturing plant.

Likewise, value analysis can show that using the product decreases the purchaser's processing cost.

New wood panel and insulated foam products help save on energy costs, thereby increasing the product's value.

3. ***Helping the customer comply with regulatory requirements.***

This has become an important product differential. Adhesive and sealant manufacturers have seized the opportunity to develop and market a wide variety of industrial products customized to meet process requirements in the wood products industry. Driven by government regulations to minimize volatile organic compound (VOC) emissions, wood products manufacturers have sought help in moving from solvent-borne products to water-borne formulations and high-solids products. The adhesive and sealant industry has provided assistance with products differentiated on the basis of problem solving and service for complying with VOC emissions.

This service orientation has earned the adhesive and sealant formulating industry some immunity from competition from imported adhesives.

4. ***The product meets health, safety and environmental concerns.***

Consumer awareness of health, safety and environment issues influences an industrial buyer's selection of a product. Compliance with current standards of acceptability becomes a product differential for industrial products used in the manufacture of consumer products.

Product safety concerns have been around for many years but have recently moved to the forefront. The manufacturer of consumer products bought by the consumer is liable for their safe use. The liability is passed on to suppliers of wood components or wood products used in manufacturing the consumer product.

Health and nutrition acceptability is now a consumer standard. These standards apply to home construction and home furnishings as well as to food and drug products. An example is the temporary setback encountered by the Permanent Wood Foundation (PWF) in the Southwest when a contractor claimed that radon accumulated in a PWF basement. Formaldehyde emissions are another health problem that provides an opportunity for differentiation.

Fletcher Wood Panels Ltd. of New Zealand is differentiating its panels by producing special lines of low formaldehyde particleboard and medium density fibreboard. They identify the low formaldehyde products with a blue cross and use the brand names Pynefloor Plus and Lakepine Plus to identify the particleboard and MDF.

Environmental acceptability is also a product differential. Environmental acceptability includes products that are recyclable, nontoxic, and use biodegradable components in processing as well as the method of manufacturing or growing the product. Detected pesticide residues already make fruits and vegetables suspect. Herbicides used in forest management have also acquired a negative image.

5. *Manufacturer's credibility*

A skeptical public is wary of advertising claims that do not live up to expectations, and the public's suspicions have been heightened by the accusations of consumer groups. A product bearing an Underwriters Laboratories seal of approval, or a stamp indicating compliance with American National Standards Institute (ANSI) or the Juvenile Products Manufacturers Association (JPMA) or similar certification is a valuable product differential.

6. ***Style trends***

Style can make a positive statement of differential advantage for a manufacturer of industrial wood products, as well as consumer products. A manufacturer of legs and arms for the furniture industry can create a positive differential advantage by furnishing parts in the modern or antique or other style in vogue.

7. ***Offering superior service***

Some differential advantages capitalize on superior service as part of the package of product benefits. Customer service has been called the business battleground of the 1990s. A study of customer expectations by Leonard L. Berry, professor of retailing and marketing studies at Texas A&M, concluded that customers want companies to do what they say they are going to do—to keep their promise of good service.

8. ***Offering favorable credit terms***

Extension of credit is a powerful product differential. To meet customers' requirements, roofing wholesalers must have roofing products available when and where roofers use them. Usually roofers depend on credit until they are paid by their customers, and they buy competitively priced material to make their bids as low as possible.

9. ***Providing technical support***

Customers look to manufacturers for other services such as: exceptional technical support, clear technical information on use of the product, state-of-the-art communication techniques for sales instructions (videotapes, instruction manuals illustrated in comic book style or interactive computer programs), the availability of a problem-solving technical service organization, the availability of technical service experts 24 hours a day seven days a week, and product salesmen trained to meet customer requirements.

10. ***Provide follow up after installation***

The service differential can be augmented by another product advantage: follow up after installation. Industrial customers want assurance that after the sale is made and the product is in use, manufacturers will continue to be interested in their welfare.

11. ***Ability to provide an immediate quotation***

Atlas-Roll-Lite Co., an Edison, N.J. maker of industrial doors, installed a simple computer system that enabled the front order desk to instantaneously price and schedule orders. Atlas also prints out a copy of the design drawing while the customer waits. The use of this program has saved customers from the exasperation of waiting a week or more for a price and delivery quotation.

12. ***Deliver products on time.***

The company that can deliver faster than anyone else in the industry has a differential advantage.

Some companies gain a competitive edge by quick response—24 hour turnaround, processing while the customer waits, 10 days from order to installation, instant gratification. Fast delivery is the strategy that made Domino Pizza a household word in a highly competitive market. Contrast the quick delivery strategy with the practice of some furniture manufacturers who keep a customer waiting four to six months for a sofa.

An adjunct to timely delivery is just-in-time delivery (JIT). JIT is both a marketing concept and a manufacturing practice. A manufacturer operating on a JIT principle organizes production so that every part arrives in working order at the right place in an assembly just in time (JIT) to be installed. JIT management has been shown to produce better quality products at greatly reduced operating costs in less production time. That's the production side. On the marketing side, gaining a reputation for delivering what the customer orders when the customer needs it is a positive product differential.

Hillenbrand Industries, Inc. of Batesville, Indiana used the JIT concept to attain the position of industry leader in the United States casket business. Supplying about one-third of all the caskets used in the United States every year, Hillenbrand offers funeral directors the opportunity to choose the desired casket from a catalogue providing 300 basic casket models, instead of maintaining an expensive inventory. A system of 59 distribution centers delivers the selected casket, usually within 48 hours. Using JIT, Hillenbrand meets specific customer needs with an impressive variety of custom caskets. Manufacturing is rigorously streamlined, allowing the company to keep its prices consistently competitive.

The Hillenbrand marketing department vigorously promotes the product differentials—speedy delivery, wide choice, competitive prices, and an array of services to funeral directors. In addition to videotaped management and merchandising programs and books to help funeral directors with their difficult and delicate job, the company plants a memorial tree in a national forest for every casket sold.

The burial casket market appears to be subject to increasing segmentation according to a 1989 government survey of funeral expendtures. The alternative of cremation has encroached on traditional funerals, especially in the West where one-third of the funerals are cremation (compared with 13 percent in the Northeast and only 10 percent in the South and Midwest.) Spending on funerals is segmented by education (the less educated are the least likely to arrange for cremation), income (the lower the income the more likely to elect the traditional burial), and racial lines (non-whites are more likely to choose casket burial).

13. ***Establishing a reputation for reliability***

Piggy backing on the respected brand names of other products, such as Congoleum, General Electric, or IBM, when possible, helps a product's reputation for reliability.

Congoleum Corporation uses its reputation to market the "Town & Country" line of wide width sheet vinyl flooring tailored for modular housing. "Your reputation opens doors, " it advertises to modeluar home builders, " Ours closes sales."

14. ***Availability of flexible specialization***

A company with flexible specialization uses artisans and craftsmen for specialized tasks. Flexible specialization allows shorter production runs, and increased collaboration between the supplier and customer. Flexible specialization is used to develop crucial components and subassemblies that can't be achieved by mass production.

15. ***Offering large discount spreads***

A product may also be differentiated by distribution practices such as offering a larger than customary discount spread.

16. ***Using a well-known distributor***

Using a distributor who has won the trust of customers can greatly enhance product differentiation.

Robinson Brick Company, Denver, Colorado used the distribution differential as a tool to transform a local commodity, bricks, into a national differentiated brand product. The 1984 recession in Colorado pushed Robinson Brick into utilizing differential product advantages it had previously ignored. They made a better brick than competitors, offered a greater range of colors, and they delivered bricks faster than anyone else in the brick industry. The company based its new differentiation strategy on those advantages, using them to convert Robinson bricks to a nationally recognized brand.

Robinson had to solve two problems to achieve the conversion, and it couldn't afford to spend much for the solutions. One problem was what to say. The second was how to get the word out. Robinson solved the first problem by playing up the fact that Robinson bricks meet the American Society for Testing Materials (ASTM) standards for facing brick extra (FBX) (they had been meeting ASTM standards since 1970 but the company had not recognized its marketing value.) Robinson bricks spelled quality.

Without the finances to spend $30,000 per page in advertising, or to obtain a celebrity endorser, Robinson Brick decided to use distribution as the key to get the word out. Setting up a national distribution system, one distributor at a time, took patience, but it was cost effective and it worked, despite the barrier of the high cost of freight for heavy bricks. Out-of-state sales increased, more than compensating for the continued slump of in-state sales as a result of the prolonged recession.

George Robinson summed up his strategy for developing a brand name to market a difficult-to-differentiate product in an interview with *Inc.* magazine, June 1990:

- ❑ Develop a better product
- ❑ Tell people why you are better
- ❑ Use word of mouth to influence the people who influence buyers
- ❑ Find an independent authority who will certify that you have a better product. (The ASTM played the role of independent authority for Robinson Bricks.)

17. ***Other differential advantages***

- ❑ Unusually favorable warranties or guarantee
- ❑ Use of the product by another well-known manufacturer
- ❑ Meeting rigorous quality standards
- ❑ Made from preferred species

Differentiating Consumer Products

The product differential that helps make the sale to the industrial customer may not get anywhere with the consumer. The perspectives of consumers products are as diverse as the people themselves.

Do consumers buy on price? Sometimes. Often consumers buy convenience goods on price, but they are willing to pay more for high quality specialty goods, such as dining room furniture.

Just under half of all American households have annual incomes below $25,000. This economic segment buys low-priced wood veneer cabinets and imitations of expensive perfumes.

Midscale households with incomes of $25,000 to $50,000 account for one-third of American households, and often have expensive upscale tastes.

Upscale households (over $50,000 annual income enjoyed by about 20 percent of the households) may buy bargain oil in a discount store for their $30,000 Porsches which they drive to the theater where they have paid $60 for the ticket.

Do consumers buy only for convenience? Sometimes. When packaged cake mixes were first introduced, homemakers were slow to put them in their grocery carts. Traditional homemakers were plagued with guilty consciences, a fear of being considered a poor homemaker if they did not start from scratch. It took a couple of decades and an exodus of housewives from full-time kitchen duty to the workplace to elevate convenient cake mixes to respectability.

Today busy homemakers like the convenience of new finishes that make caring for a wood table top as easy as wiping off a plastic top. Laminated products that combine the elegance and ambiance of wood with the convenience of plastic surfaces are quickly gaining popularity. Combining convenience and the feel of wood makes a powerful product differential.

Do consumers buy on quality? Often, when they recognize the difference between good quality and poor quality.

Quality is a persuasive product differential, offering customer satisfaction and an opportunity for effective promotion. But relatively few customers understand what quality is.

In purchasing upholstered furniture, consumers have been found to be aware of design and color, but generally uninformed about the quality elements of upholstery. If cushions feel wrong, customers are dissatisfied. They rarely know anything about the foam density that gives a better feel.

> England Upholstery Manufacturing Company, a manufacturer of promotionally-priced upholstered furniture, solved the costly problem of consumer dissatisfaction and saved money by upgrading the quality of the foam cushioning used in its products. Foam cushioning with a 1.0 pound per cubic foot density (pcf) foam is standard in furniture priced comparably with England's. England Upholstery increased foam density to 1.8 pcf, a density typical of higher priced furniture. Foam quality became the key product differential for the company. The reduction in cushion returns more than offset the additional cost of the better quality foam.
>
> To market the extra value of the upholstered products, England created another differential advantage. It instituted a warranty program: frames were warranted for 20 years, fabrics for two years, foam cushioning for five years. This enhanced England's image both with the consumer and the retailer. Their product differentiation program has been responsible for a company growth of 40 percent a year!

Growing consumer concern for the environment has created a new array of product differentials. Manufacturers can differentiate products by:

- ❑ Emphasizing environmental safety
- ❑ Demonstrating social responsibility
- ❑ Recognizing nutrition concerns
- ❑ Acknowledging humane concerns

Celestial Seasonings, the company known for herbal teas, was pressed by its customers to discontinue using chlorine bleached bags for its teas. Some tea drinkers worried that pulp mills create dioxin when they use chlorine bleach. Now, one of Celestial Seasoning's product differentiations is its use of oxygen bleached paper.

Proctor & Gamble responded to landfill concerns by differentiating its disposable Pamper diapers, using reduced cubic volume to take up less landfill space and a new recycling system to separate the plastic for recycling and the pulp for further use.

Duncan Hines differentiates its muffin mix with the label: "No preservatives, no palm oil, no cholesterol."

StarKist Tuna bears a label stating that it is "Dolphin safe," and some cosmetics bear emblems showing that they are not tested on animals.

Differentiating Niche Products

Niches are carved out by manufacturing and marketing products for specialized uses. Niche products are readily differentiated from other products.

The experience of Taylor Mattress Co. is a model for manufacturers of niche wood products. The Salisbury, North Carolina business which has been in the family for generations, found that its mobile home and recreational vehicle market was so good that it attracted cutthroat competition. The company switched to contract work with jobbers, which brought with it a new set of problems.

Taylor finally found its niche in customized and high-end bedding. Now distributing mattresses direct to the factory, Taylor President John Heilig has found the specialty market more interesting and innovative. "I don't ever intend to gear up and see how much I can sell," Heilig said, adding that he'll never make a full range of traditional bedding again.

Some of the most successful niche products are premium items for upscale consumers. Premium products are typically of superior quality, distributed through highest quality channels, and advertised with discretion. To consumers of premium products, price is no object. They are affordable indulgences such as Haagen Dazs ice cream or dijon mustard. They may be items that demonstrate the buyer's superior taste like Coach Handbags or Pierre Cardin fragrances. Other products are premium because they are considered smart investments, like Maytag washers and dryers, known for their reliability. Acquisition of a premium product is a status symbol conveying the impression that the owner is wealthier, more educated, more cultured, or in some other way better than the average person. Mercedes Benz and Cadillac cars are in this category.

Premium products constitute—at the most—20 percent of the market. Although fewer in number than middle and low income segments, these conspicuous consumers are a good market. So good that many business economists are advocating that American companies focus on higher income customers. The high volume, low cost production approach, they reason, is vulnerable to foreign competition.

The premium end of the market is less vulnerable to price competition. Mainstream products are squeezed between low cost producers and premium quality producers. Some American furniture manufacturers and textile producers have looked at upscale customers and found the better profit margins in this market worth pursuing.

Product differentials for premium items influence the purchasing decisions of upscale consumers in several ways. One way is to feature technical superiority. Maytag washers, for example, require fewer repairs, therefore, they are technically superior.

Featuring the prestige image where the benefits to self-image sell the product is nothing new.

> The 1895 Montgomery Ward Catalog furniture section includes item 72720, a Cheval Chamber Suit (bedroom suit). "If you buy this suit," the catalog promises, "your neighbors will all envy you." (Most of the catalog furniture was made of solid oak, quarter-sawn oak, or hardwood.)

> Perrier water may be the ultimate in a differentiated product. It is a special water for special people. This image has been reinforced by communicating the product's heritage and place of origin, making it "nature's soft drink." This type of product differential is similar to the singularization of wood products.

Featuring uniqueness is an irresistible premium product differential. When the product's functionality is not an issue, technical superiority, with its uniform results, may be a detriment.

For example, machine made glassware is uniform, without irregularities. Hand blown glassware is more likely to have irregularities. But these flaws not only do not impair the function of the product, they are also visual indicators for all to see that the glassware is hand crafted and therefore unique and rightfully more expensive.

Character wood panels have the same appeal. The flaws the manufacturer sees as defects in quality lend individuality and character to wood panels. They are priced at the higher end to fulfill their function of bolstering self esteem and pride of ownership.

Difficult, scarce, unusual—these attributes almost guarantee that an item will carry prestige. Hand crafted items take time to make. And some are difficult to make.

> The Quinault Natural Resources, a magazine published by the Quinault Indian Reservation in Western Washington, described *Ohia*, a native premium Hawaiian timber species, in its Summer/Fall 1988 issue as "an unusual-

> ly hard wood—hard and weighty. It is used by the Navy as flooring for its railroad cars and for piers. Wealthy Hawaiians use it for cabinets and floors because it is so hard. It is also difficult for a cabinet maker to lathe." Difficult, therefore expensive, and so a status symbol.

Unless the purchase of the product feeds the psyche in some way, newness alone does not result in customers beating a path to the manufacturer's door. Innovation in itself is not an effective differential for a premium product.

When the television was first introduced in the late 1940s and early 1950s, high cost and limited program offers turned away even the prestige buyers. As the screens became larger and the number of programs available increased, television rose quickly from a premium product for the very few to a mainstream item. The advent of color television was likewise slow to catch the public fancy, perhaps because programs in color were limited.

Marketing premium products requires total commitment to the quality image. Performance warranties, expensive packaging, and brand logos that connote quality are used to enhance a quality image. Many marketers of premium products carry the quality image through ***forward integration*** by establishing specialty stores to sell their line. Products like Polo Fashions and Haagen-Dazs ice cream are retailed in conjunction with life style concepts. Some premium products enhance their quality image by association. One TV commercial plays opera music against a background of elegant crystal and expensive table settings to indicate the quality of the product it advertises.

The concept of singularization is a product differential that moves ordinary wood products into the premium class. A simple breadboard moves into the class of a prized hostess gift if the wood grain is highlighted and it is elegantly packaged, using a slogan and a description of the unique character of the wood from which it is crafted. Paneling, moulding, furniture, cabinets, almost any item made of wood, can become a premium product if it is so differentiated.

The producers of premium coffees are experts at product differentiation. All coffees aren't the same, anymore than all woods are the same. The arabica beans are known by coffee lovers for their full, rich flavor. Consumers of gourmet coffee willingly pay $8 or more for a pound of these beans. Mass producers of coffee generally use the low grade robusta bean, which sometimes produces an acidic, bitter flavor. Starbucks Coffee Co. bars, known for their gourmet coffees, translates this $8 pound of beans into the equivalent of $20 per pound of brewed coffee sold by the cup. Despite the decline in coffee consumption in the past thirty years, from an average 3.1 cups a day to

1.75 cups a day, sales of the better grades and roasts of coffee have doubled in the past five years! Another testimonial to the value of singularization!

DEVELOPING PRODUCT DIFFERENTIALS

The wants and needs of potential customers often are not reflected in the design and production of the final product. Marketing and engineering, even within the same company, speak different languages and have difficulty talking to each other. When they do communicate, customer satisfaction is usually increased.

Japanese manufacturers have developed the science of getting marketing and engineering to work together in designing products. The communication and coordination of marketing and engineering has become the basic design tool of a management approach called the "house of quality."

This management concept, devised in 1972 by Mitsubishi, is a set of planning and communication routines that becomes a kind of map for coordinating design, manufacturing, and marketing. The foundation of the house of quality is marketing. It is based on the principle that products should be designed to reflect what customers want, rather than what's convenient to manufacture. It begins with customers describing, in their own words, the products and product characteristics they prefer. An example of customer attributes for the house of quality in designing an automobile is shown in Figure V-2. What customers want in a car door is summarized in almost 20 desires from "easy to close" to "easy to clean." It ends with engineers and designers creating a product that meets those desires.

Sounds simple? The concept is. What is not so easy is getting people thinking and working together in a house of quality organization.

The house of quality may sound like something that fits only larger companies. But smaller, less bureaucratic companies may be able to adapt the concept even better. In large companies, marketing people have their domain, designers theirs, and often management is out of touch with both. In smaller companies most executives wear several hats, and the close communication between designing or engineering, production, and marketing for the house of quality can take place during a coffee break.

Customers mean different things when they say "quality," and it is a challenge to design products that satisfy all of these quality ideas at the same time. Langer Associates, a New York based marketing research firm, was asked by the Gillette Company to conduct market research on what "quality" really means to consumers. Langer found that today's

FIGURE V-2
The House of Quality—What Customers Want in a Car Door

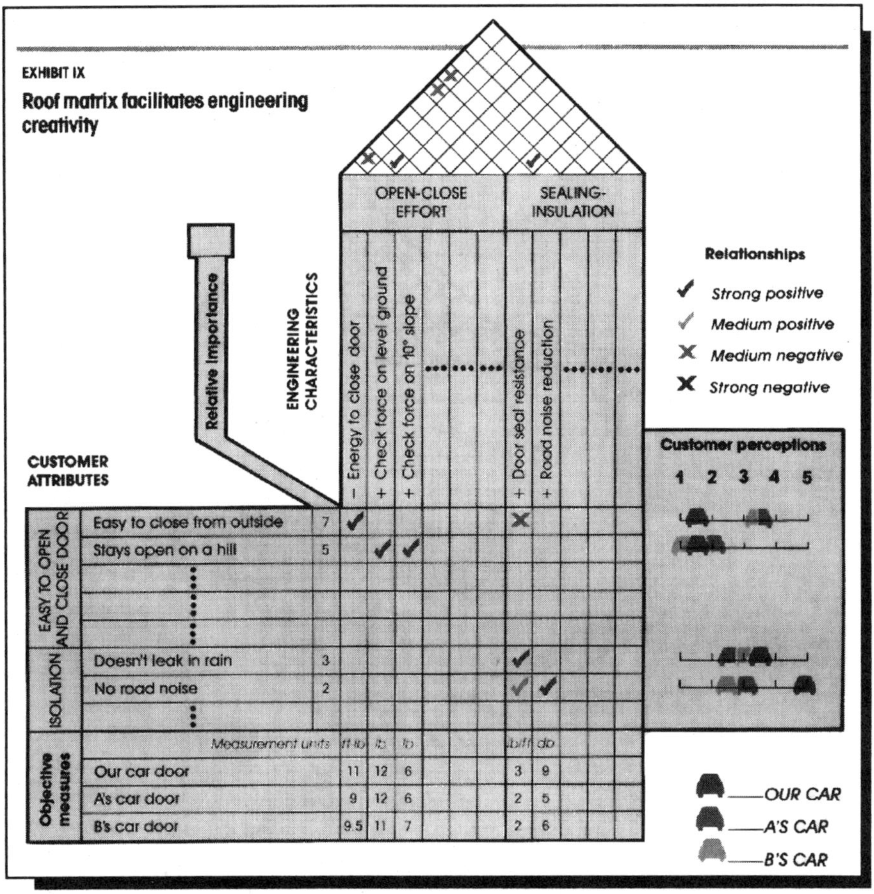

consumers are more demanding and educated about quality than in the past. They are more aware of what goes into a product, the ingredients in food, the cloth for clothing, and the wood in home furnishings.

A poll taken by the market research firm of Yankelovich Clancy Shulman taken in 1990 found that Americans rank quality components in this order:

1. Reliability
2. Durability
3. Easy maintenance
4. Ease of Use
5. A known and trusted brand name
6. A low price

Five years before a Gallup poll reported that consumers ranked quality components differently.

1. Well-known brand name
2. Workmanship
3. Price

Furniture designers search for patterns, color combinations, textures, and styles to set their lines apart from others. The tapestry and style differentials are sometimes emphasized, while the wood frame is largely downplayed.

> An item in the May 1990 *Upholstery Design & Manufacturing* reveals the size of the gap between designer's knowledge and the general public's. The item reports on an 18th century Queen Anne settee reproduction produced by Century Furniture, Hickory, North Carolina. The report described in detailed decorator terminology the origin of the tapestry and the chair details, but the wood frame is passed over lightly with the note that the legs were machine carved in mahogany and the doweled, glued, and screwed frame was made of maple and poplar. No singularization there!

Appealing brand names, packaging, and product color differentiate products from each other. Brand names influence product sales, sometimes in unexpected ways.

> Toro Corporation originally named its lightweight, efficient snow thrower, "Snow Pup." When sales of the snow thrower failed to meet projections, market research revealed that the brand name, not the product, was the problem. "Pup" gave consumers the impression that the snow thrower was not powerful enough. When Toro changed the brand name of the same snow thrower to "Snow Master," sales increased significantly.

> Matex Corporation also found that a name makes a difference in sales. They had named their superior rust proofing compound Thixo-Tex, a name which conveyed little to the buyer. A change of name to Rusty Jones, identified by a trade character, helped increase sales from $2 million to $100 million in four years.

Packaging should also reinforce the differential advantage. Busy, bright wrappings may detract from a quality image.

Product color may also make a difference in sales. This is especially important for cabinet makers and furniture manufacturers as colors go in and out of fashion. Light birch was the color of the 1950s, dark walnut won customers in the 1960s, oak headed the popularity contest in the 1970s and 1980s. Light, bright colors appear to be making a comeback in the 1990s, but manufacturers might think twice before featuring red cabinets.

DIFFERENTIATED MARKETING OF A PRODUCT LINE

Differentiated marketing strategies are used when a company markets a number of products, each designed for individual segments of the total market. The objective is to increase total sales and develop loyalty to products in each part of the market.

The automobile industry is often used as an example. Ford offers five types of cars, each appealing to a segment of the new car market. Proctor & Gamble markets at least eight different detergents, each appealing to a segment of detergent buyers. The manufacturer of kitchen cabinets may market both traditional and frameless cabinets, each appealing to a particular group of buyers. Some firms practice concentrated marketing, focusing their entire effort on satisfying a smaller market. The automobile industry again furnishes an example. Volkswagen of America for 20 years concentrated on buyers looking for an economical and no-frills practical car. Fisher-Price, on the other hand, concentrates on high quality toys for the upscale children's market.

WHEN THE PRODUCT DIFFERENTIAL IS NEGATIVE

Product differentials are assumed to be positive. What happens when the differential is negative? When the brand name or company name or method of manufacture becomes tarnished? What does the marketer do when an otherwise good product becomes the victim of a negative reputation?

Handling negative product differentials requires skill. A negative differential cannot be talked away. It must be overcome, and differential benefits must be created to overweigh the negatives. If there is no public consensus that the differential is negative, political science techniques come into play.

Today, businesses are discovering that consumer concern with environmental and social responsibility is affecting marketing. The December 3, 1990 *Fortune* magazine addressed "How to Deal With Tougher Customers." *Fortune* points to a new era in marketing every product, including wood products, with these conclusions:

- Americans are willing to pay a premium for quality
- Americans are concerned about nutrition and want to know what's in their food
- Americans worry that their disposable society is despoiling their environment
- Americans no longer trust manufacturers

Today's major buyers, the maturing baby boom consumers, are informed, articulate, and have strong opinions. Their children may be expected to reflect the opinions of their parents who, according to *Fortune*, are "leaner, meaner, more cynical" than their grandparents.

The green movement also affects product acceptability. So powerful has the movement become that now seminars on "The Power of Green" attract top industry executives. The executives learn about consumer attitudes toward green products, using environmental concerns to increase competitiveness in global markets, and how to assure that products and packages pass environmental litmus tests.

Furniture manufacturers and designers around the world are running head-on into strong environmental movements against using endangered wood species. Environmentalists are pressuring wood products consumers and manufacturers to avoid using endangered species and those from tropical rain forests. The 220 local offices of Friends of the Earth in Great Britain want to control woods that can be used in furniture. A leading architectural firm in Britain no longer specifies tropical hardwoods such as iroko, ramin, mahogany, and teak (although Burmese teak, a managed species grown on plantations, appears to be ecologically acceptable to some English consumers). A U.S. designer who used Honduras Mahogany was criticized in the English press.

Germany and Switzerland are already banning tropical timbers in government buildings and the European Common Market 1992 is expected to participate in pressuring against tropical imports.

The movement has spread to the United States. The Rain Forest Action Network and the Rain Forest Alliance protest against the use of tropical rain forest woods have used boycotts and demonstrations to attract public attention to their view. Congress has considered legislation against importing tropical timber, including bills prohibiting the import of teak from Burma, labeling wood imports showing the country of origin and the scientific species name.

The issue of tropical timber directly affects about 200 million people who make their living from the tropical rain forests. At least 15 developing countries depend on the exports of their tropical hardwoods for foreign exchange, which is said to be worth more than $8 billion annually.

The forest industry has not made a convincing case for utilization of tropical rain forest species, although logging has been evaluated as the least important factor in the destruction of the tropical rain forests. Slashing and burning to raise crops and beef are said to be the major causes, followed by cutting for firewood and denuding in preparation for mining.

The negative differential caused by concern over tropical rain forests is too powerful to be ignored.

Negative product differentials may develop quickly but they change slowly:

Exxon is a well known brand name, but after the Exxon *Valdez* oil spill, the company's image slipped into disrepute. People began tearing up their Exxon gas credit cards. Consultants in public opinion generally concede that Exxon bungled in handling the negative image.

Exxon's poor performance in coping with its negative differential contrasts with Johnson & Johnson's finesse when some Tylenol was found to be poisoned many years ago. Although the poison incidents were few, Johnson & Johnson immediately recalled all Tylenol from the market, showing that they cared more about the public welfare than immediate profits. The Tylenol brand name remained virtually intact.

Foam cups and packaging, disposable diapers, and plastic products have been tarnished as contributing to the garbage glut. Retailers and fast food outlets are swiftly changing to biodegradable products, and the foam packaging and disposable diaper manufacturers are responding that their products are now biodegradable. Some retailers are turning environmental concerns into positive differential advantages.

Wal-Mart Stores, Inc. appeals to customer concerns about overflowing landfills by advertising that: "We're looking for Quality Products That Are Guaranteed Not to Last," implying that the products or packaging will degrade in the landfills. Wal-Mart also applies a "Wal-Mart Commends" tag on the products from manufacturers who claim they have altered a product to help the environment.

Macy's and Lord & Taylor are among department stores attempting to save the African elephant by proclaiming that ivory is out of fashion.

Objection to the practice of harvesting timber by clearcutting forests led the continuing parade of environmental grievances against the forest industry. The era of benign public neglect that the forest industry had enjoyed for some years ended with the passage of set-aside legislation.

The forest industry has responded to the environmental movement with defensive advertising and educational programs. It's still finding it difficult to handle the negative differential. The standard response to any measure that decreases the availability of timber is the number of jobs lost and the area's decline in revenue. These calamities have not impressed the public.

Concern about the environment has resulted in wilderness set-asides. The Endangered Species Act and legislative and judicial action have further reduced the availability of the raw material for the forest industry.

Animal rights groups have vigorously attacked the fur industry, and some analysts say that actions by such groups as People for the Ethical Treatment of Animals (PETA) have caused fur sales and profits to decline. The industry appears to have made little effort to handle this negative product differential, blaming flat sales on economic conditions and changes in fashion, rather than anti-fur activism. Manny Mohl, president of the New York fur maker, Mohl Fur Company, sums up the industry's answer to anti-fur activism this way: "Show a woman a nice mink, and she'll never turn you down." The upholstered furniture industry is responding to the negative product differential associated with the use of chlorinated fluorocarbons (CFCs) in upholstery foams by moving away from their use as quickly as alternatives develop. These ozone depleting chemicals, which are said to contribute to global warming, have been under attack for some years. A recent federal law places an excise tax on any business that manufactures, produces, or imports CFCs.

Entrepreneurs have used the wave of environmental activism as an opportunity to introduce new products, differentiated on the basis of environmental acceptability.

> The Plastic Lumber Co., Akron, Ohio, assumes that the products it manufactures from recycled plastics provide a positive differential benefit to the environment. They further assume that this benefit to the environment will induce consumers to pay more for picnic tables, mailbox posts, and speed bumps made from recycled plastics than they would for similar wood and concrete products.
>
> Plastic lumber can be sawed, nailed, drilled, glued, and bolted like wood. It is lighter, requires less maintenance, and is said to be impervious to the weather. But there's a catch: recycled plastic items cost up to twice as much as the same items in wood or concrete. Despite repeated polls indicating that consumers say they will evaluate the environmental impact of products in their buying decisions, the public hasn't beat a path to The Plastic Lumber Co's door.
>
> As Nancy Pfund, partner in Hambrecht & Quist, a San Francisco venture capital firm, says: "It's a common mistake to think people are going to buy products just because they are environmentally correct. You can't rely on that kind of altruism."

When the Negative Differential is Personal Risk

Formaldehyde emissions. Wood preservatives. Dioxin from pulp mills. Radon build-up in permanent wood foundations (PWF). Fear!

When people are afraid that they are risking their lives when they use a forest product, a negative differential is created. Never mind that the actual risk may be minuscule, that many reputable scientists maintain that there is no risk. Once the fear of personal risk is introduced, it becomes an overwhelming differential disadvantage.

Coping with the consumer's fear of personal risk is probably the most difficult challenge in marketing. Communicating to consumers that their fears are exaggerated collides head on with the quirks of the human mind.

Decision Research, a nationally renown research firm specializing in communicating with the public about risk, has discovered five factors that affect how the public views risk of personal safety. Application of their findings should help forest industry marketers handle fears of personal risk:

- People's perceptions of risk are often inaccurate, and are influenced by memorable past events and what they imagine future events will be. The risks in dramatic causes of death, such as cancer and natural disasters, tend to be overestimated. Risks from undramatic causes, such as asthma and diabetes, are usually underestimated.

- Merely mentioning risk frightens and confuses the public. Even the most rational discussion of possible bad effects from using a product makes the risk more real and frightening. "Worst case scenarios" required in environmental impact statements can cause very negative reactions, no matter how unlikely the worst case is.

- Strong beliefs are hard to modify. Polarized opinions about hazards such as herbicides, nuclear power, and the disappearance of ancient old growth forests reflect people's reactions to the gambles of life. Industries, including the nuclear and forest industries, spend millions to inform and educate the public in the hopes of modifying these polarized positions. The effort is often in vain. According to psychological research, once a belief has entered people's minds, it persists stubbornly, even in the face of contrary evidence.

- Naive views are easily manipulated by the words used to present information about personal risk. The wording itself can determine people's reaction to those risks. Statistics framed in terms of dying produce negative reactions, the same statistics in terms of surviving produce positive choices.

- It is difficult to educate people about risks. Their apprehensions can often be traced to biased experiences, which may be countered by information presented in positive terms.

Industries often fail to ease fears about personal risk because they assume that they and the public share a common perception about a risk. Risk means different things to different people. People's behavior in face of risk is influenced by whether they view the risk as controllable, uncertain, or with the potential to be catastrophic, inequitable, or a threat to future generations.

What do these research findings mean to the plywood industry in coping with reports of chronic illness due to formaldehyde emissions? Or to the paper mill attempting to obtain a permit for expansion while the public worries about dioxin in its streams?

They mean that an industry or company is unlikely to win public support by comparing the risk of its product with the statistics about other risks. Toxic probabilities, annual mortality rates, or loss of years of life expectancy are meaningless to most people. Show the public, rather, that formaldehyde emissions can be controlled with reasonable certainty. Minimize the perception of catastrophic potential with a warranty or low-cost insurance policy.

Some companies believe strongly that if they don't mention the risk people won't think about it. That belief is easily undermined by media coverage. Acknowledging the risk and winning credibility by being honest and forthright about protecting the public has been found to be the most effective strategy for handling negative differentials.

POSITIONING

In marketing, the differential advantages are the benefits that distinguish one product from another. Promotion communicates those differential advantages to potential customers. Positioning aims to place the product benefits so firmly in the customer's mind that he or she thinks first about that product when making a purchase.

Xerox, IBM, and the Coca-Cola Company have successfully positioned their products. Their products are the first to come to mind when a customer thinks about a purchase. Positioning is the marketing strategy aimed at establishing a product in the minds of prospective customers.

People say "make a Xerox" when they mean make a photocopy. They say they are using "an IBM" personal computer when they may be using any one of many different IBM clones. They ask for a "Coke" when they mean another soft drink.

The objectives of manufacturers, distributors and retailers meet in positioning. The manufacturer has designed and made a product that uniquely fills customer needs and expectations; the distributor facilitates getting that product to the market segment that needs and wants what the product has to offer. The retailer prefers to stock products that turn over quickly.

Positioning brings together the total marketing effort. It cements the image of a product that will satisfy the customer's interest and provides evidence of the firm's marketing intentions.

Effective positioning is based on real benefits that distinguish products from competitors. Clever slogans and empty promises do not win product preference in the market.

Positioning has become a major strategy in a society bombarded with messages urging consumers to buy products or adopt a life style or champion a cause. In their book, "Positioning: The Battle for your Mind" (New York, McGraw-Hill, 1981) Al Ries and Jack Trout shed light on the role of product positioning:

> " . . . America consumes 57 percent of the world's advertising . . . . The average American family watches television over seven hours a day. An eight

ounce package of Total breakfast cereal contains 1,268 words of copy on the box . . . . The average supermarket in the United States has some 10,000 individual products . . . on display . . . for a market where the typical new high school graduate has a working vocabulary of about 18,000 words."

Add to these statistics other circumstances that distinguish today's customers and the significance of positioning in current marketing is apparent:

- ❑ Customers are busier. Since 1973, the average work week (including commute time) has increased by six hours.
- ❑ Families have less time for shopping. The homemaker mother has turned into a harried working mother who retains her old responsibilities in addition to an outside job. Men also have less disposable time.
- ❑ The average American has 37 percent less leisure time than in 1973. People are taking shorter vacations, often four day weekends.
- ❑ American consumers report that they want better quality than they have received previously.

Product positioning aims to help these busy customers work through the bombardment of messages to select specific products.

The decisions a manufacturer makes on positioning influence all facets of the business—promotion, packaging, channels of distribution, price, and manufacturing.

Positioning: the Moment of Truth in Marketing

If a retailer does not stock a product, the money poured into product development and promotion may be wasted. Without products, on the other hand, the retailer has little to sell. Manufacturers and retailers depend on each other, an interdependence often ignored.

The link between manufacturers and retailers is increasingly important. Nowhere is this more evident than in home improvement stores. The growth of these retailers has increased the opportunities of the wood products industry to reach segments of the consumer market not before available. The home center stores—defined as any operation with more than $500,000 annual sales that sell lumber, building materials, hardware and home improvement products—are a world removed from the traditional hardware store and lumber yard.

Traditional lumber yards—operations with lumber and building materials businesses greater than 60 percent of sales—typically make more than half their sales to contractors.

FIGURE V-3
Wood-based Items Stocked by Home Center Stores, Hardware Stores, and Lumber Yards

	Percentage of Stores that Stock and Sell Each Item		
Item	Home Centers	Hardware Stores	Lumberyards
RTA Furniture			
Finished	22%	54%	1%
Fencing	82%	4%	11%
Hardwood			
Flooring	38%	49%	9%
Entry Doors	90%	3%	67%
Hardwoods	63%	1%	38%
Treated Lumber	92%	2%	37%
Moulding	95%	55%	43%
Wall Paneling			
Solid Wood			
Pre-finished	80%	2%	7%
Plywood	96%	3%	34%

Source: Home Improvement Center November 1990

In contrast, lumber and building materials play a minor role in hardware stores—defined as operations with paint, plumbing, electrical, and hardware business greater than 60 percent of sales.

In a 1990 survey of almost 7,000 home improvement centers (including home centers, hardware stores, and lumber yards) *Home Improvement Center* magazine found a number of differences shown in what each type of store stocks and sells. (See Figure V-3.)

The way home improvement stores position products can guide wood products manufacturers to expanding sales. These retailers succeed by giving customers what they want. The manufacturer who provides the products prospers by helping the retailers succeed. With 92 percent of all home centers stocking treated lumber, manufacturers of this product will benefit by observing how they position the product.

Figure V-4 summarizes suggestions from retailers on how manufacturers can do a better job of helping them sell more home improvement products.

FIGURE V-4

RetailersTell How Manufacturers Can Do a Better Job of Helping Them Sell More Home Improvement Products

In a 1990 survey, *Home Improvement Center* magazine asked retailers: "In your opinion, how can manufacturers do a better job of helping your firm sell more home improvement products?" Retailers offered these suggestions for positioning products:

Suggestions	**Percent of retailers suggesting**
More advertising	24%
Better POP displays	22%
More informative packaging	18%
Better pricing	12%
Better quality control	11%

Fewer than 10 percent of the retailers suggested more store training, smaller minimum purchases, shorter lead times on orders, better product warranties, and more knowledgeable sales representatives.

Further comments from retailers are instructive:

On advertising: ". . . the better educated the consumer, the easier it is to sell . . . . Help us educate consumers with national advertising and easily understood literature we can use with the sale. . . . Provide more videotaped product sales training. . . . More cooperation with in store merchandising is a must. We need to be able to work with manufacturers, not just sell their products."

On POP displays and packaging: "More point of purchase (POP) displays and packaging that state the advantages offered by the . . . product . . . Better packaging, displays, and handout literature. More information . . . Offer more POP displays and other marketing aids . . . Make packaging either with see-through or more color graphics . . . for the consumer to figure out . . . Better graphics on packages. More useful instructions."

On installation instructions: "Put more installation instructions on the outside of the package . . . so the consumer can see if he or she has the skills to accomplish the project. . . . Give easy to understand instructions on packages with a toll free 800 telephone number. . . . Labels or instruction sheets with . . . 'tools needed for this installation' and 'other parts needed' . . . Make do-it-yourself products easy to install, idiot proof if possible."

On deliveries: "In today's world of jets and rockets, 12 weeks delivery is a lifetime and a burial for many quotes . . . Get it to me ASAP."

Source: Home Improvement Center November 1990, p.51

Devising the Positioning Strategy

Few products stand alone in the marketplace. A customer almost always has options about products and if none of the options appeal, a customer can choose not to buy. Customers consciously or subconsciously compare products. Positioning puts the product's best foot forward, silently nudging. The first step in positioning is selecting a product's differential advantages. The second step is identifying where competitors are weak. The third step is using the best techniques to communicate to the buyer that the product does a better job of filling his needs than competitive products.

First Step: Differentiation and Positioning

Although products have different advantages, it is often hard for buyers to detect these differences. Developing these differences in the customer's mind is the function of positioning.

Positioning strategy reflects a company's ability to understand how its product—its package of benefits—solves a customer's problems. Positioning includes customer assistance from design and application to timely delivery, warranties and return policies, service, and a supporting distribution network.

Marketing expert, Yoram Wind, writing in *Wharton Magazine, (3: 1980),* described seven basic ways to position a product, which are summarized in this section.

1. ***By a special product feature—Aspirin is positioned as an aid in preventing heart attacks.***

Armstrong World Industries' successful positioning of its sheet vinyl floor covering in the do-it-yourself market illustrates this positioning strategy. Armstrong initiated its program with market research which revealed that although women make the decision to install new floor covering, men do most of the installation. Further, most of the people who didn't install the new floor covering walked away from the project for fear of spoiling the job. More specifically, for fear of making that first cut. To combat this fear, Armstrong adopted a psychological positioning strategy featuring the ease of using the product. "Go on, cut. You'll be brilliant. Armstrong guarantees it." was the clear and focused slogan of the positioning campaign.

The strategy was supported by:

- ❑ A "Trim and Fit" kit
- ❑ A fail safe guarantee—Armstrong guaranteed that it would replace the flooring at no cost if the do-it-yourself installer made a mistake.
- ❑ Point of sale banners
- ❑ Permanent and temporary displays
- ❑ A toll free number for installation tips
- ❑ Advertising that clearly transmitted the positioning message

2. ***By a benefit—Gatorade is known as a way to replenish body fluids.***

In an advertisement headed "A Letter to Customers of Boise Cascade," Boise Cascade positioned its product line against competitive products by promising customers a total quality commitment (TQC). Boise Cascade said TQC assures "that we all have a stake in each other's performance . . . accepting less than you thought was agreed upon means you must either pass along a substandard product or take on the avoidable task of correcting it . . . It doesn't have to be that way. With TQC, it won't be . . . "

3. ***By use—Lipton Onion Soup Mix is also used for a dip.***

4. ***By user category—The secret brand of deodorant is for women.***

A prefabricated home manufacturer positions its homes to appeal to first time home buyers with the slogan "your affordable home." Another manufacturer positions its prefabricated homes for customers who are moving up into a higher quality home with more amenities by promoting "quality you deserve."

5. ***By comparing with another product—Tylenol is compared with Bayer Aspirin.***

The manufacturer of prefabricated homes has the opportunity to position a higher-end modular home directly against a site-built home. This positioning tactic is effective in attracting customers who have been disenchanted with the price and quality of site built homes. Domtar Decorative Panels positioned its thermally fused melamine panels with a "relationship strategy" designed to make the company the preferred supplier of melamine panels. The relationship strategy goes beyond price and delivery advantages to make the phrase "customer driven" a reality rather than just a buzzword. Domtar prepared for the new relationship positioning strategy by expanding its research and development, increasing production capabilities, and more than doubling its distribution network.

6. *Product class dissociation—Cherry Cola is positioned as entirely different from Coca-Cola.*

Composite panel manufacturers are positioning products by dissociating them from solid wood products:

Panfibre dissociates its medium density fiberboard (MDF) from solid wood by claiming that it is more uniform and smoother than solid wood, and available in desired sizes and quantities. Georgia-Pacific dissociates its new composite wood trim, PrimeTrim™, from solid wood trim with claims of advantages over conventional lumber trim in modular and manufactured housing. "Unlike traditional wood products," GP says, "the new trim has no knots, knot holes, wane or pitch pockets . . . resists warping, cupping, twisting, splitting, surface checking, and paint peeling." Describing the manufacturing process, the company explains that the wood is broken down into fibers, which are then reassembled in a process that "negates the imperfections common to natural wood." The list of uses for PrimeTrim™ is extensive, "anywhere defect free, high quality, factory primed, nominal 1-inch or nominal 3/4-inch-thick wood trim is needed."

7. *Hybrid usage—Baby shampoo can be used for both children and adults.*

Second Step: Positioning Against Competition

Knowing competitive strengths and weaknesses is essential in positioning. A manufacturer of hardwood flooring aimed at the remodel and do-it-yourself markets, for example, wants to know:

- ❑ Who its competitors are
- ❑ In which segments of the market do each compete
- ❑ Their competitive strategies
- ❑ Their positioning tactics
- ❑ How successful their approaches are
- ❑ Which competitors are growing, which are standing still
- ❑ Which competitors are strong enough to retaliate when a newcomer enters the competition for customers? Are they likely to lower prices? Coerce wholesalers? increase advertising?
- ❑ Are competitors using a "push" or "pull" strategy with their distributors? An established manufacturer usually can count on distributor

support when introducing a new product. The manufacturer and distributor "push" the product. But a new entry in manufacturing will not enjoy the same support. Usually the new entry bears the additional expense of promoting directly to customers to create a demand, or "pull" to encourage distributors to handle the product.

Third Step: Selecting Positioning Techniques

Positioning techniques consider how customers make buying decisions. Three views on consumer purchasing behavior are currently capturing the attention of marketing managers.

One views customers' purchases as a sequence of orderly decision making in an effort to solve problems. The customer recognizes the problem, searches for information, evaluates alternatives, and finally makes the choice. Reevaluating the choice after the purchase influences whether the customer will buy the product again.

Marketing industrial wood products assumes that the customer's management personnel involved in purchasing use orderly decision making behavior. The manufacturer helps the buyer recognize the problem, provides technical information, emphasizes the differential benefits of the product for the customer to compare alternatives, and is at the right place at the right time to take the order. Follow up and service help keep the customer.

A second view, the *experiential perspective*, emphasizes the feelings and emotions that lie behind buying decisions. The influence of belief, feelings, emotions, or other factors at the moment of decision is not an academic exercise. A significant factor in positioning is creating a belief and then reinforcing it. The role of beliefs and feelings further supports the concept of singularization. Emphasizing the individual character of each wood species and the differences within species grown under different environmental and geographical conditions creates beliefs and feelings about wood products.

The role of feelings and beliefs in purchasing behavior is complex. A consumer may use an orderly decision making process in purchasing a package of lumber to construct a bookcase on a Saturday. But feelings or emotions may also be the motivation to buy ponderosa pine or oak. These feelings may be based on a belief that the pine is easier to work, or that the hardwood is more durable.

A survey of ponderosa pine users for a New Mexico manufacturer of shop lumber disclosed that many buyers have distinct feelings about ponderosa pine, and feel that geographical origin affects the working properties of the pine.

Feelings and beliefs are also reflected in consumer reactance and impulse buying, and the tendency of consumers to spontaneously buy a new brand even if satisfied with the brands purchased previously.

Consumer reactance, according to some market experts, was the phenomenon that caused the outcry when the Coca-Cola Company changed the taste of Coke in 1985. Taste tests had shown that people liked the "new" Coke. The Coca-Cola Company was unprepared for the negative public reaction. Over 40,000 letters bombarded the company expressing customer outrage over the change. Lawsuits were filed. The story made media headlines. Why did the public react with such emotion? One would have thought that people could just switch to another soft drink if they didn't like the new Coke. An explanation for the unexpected reaction is that consumers felt that their freedom of choice had been violated. They resented Coca-Cola's audacity in changing their drink.

Beliefs and feelings influence impulse buying in which the customer buys without a prior intent to buy. It has been estimated that as many as 39 percent of department store purchases and 67 percent of grocery store purchases may be bought on impulse.

A third perspective on consumer buying is behavioral influence. Behavioral influence emphasizes the influence of building design, seating arrangements, music and other environmental factors on purchasing.

Psychological positioning recognizes the influence of a strong and recognizable product identity. Psychological positioning uses advertising, public relations, point of sale, and other communication processes to create and reinforce the product identity.

Five well established techniques are used to create a strong and recognizable product identity:

❑ Packaging

Every package is a five second commercial, according to some package designers. Emphasizing the role packaging now plays in marketing products and services, the package has been called the manufacturer's last chance at the consumer and the last five seconds of marketing. Many smaller companies are finding package redesign less expensive in positioning products than advertising campaigns. Its effectiveness in delivering the positioning message has helped packaging grow to a $50 billion dollar annual business. The forest products industry is accustomed to thinking of packaging in terms of ease of delivery (strapping lumber) or protection. But forest products, even lumber, can be packaged to improve positioning.

Home improvement stores have led the way with shrink wrapped lumber, precut lumber for fences, decks packaged in kits, and RTA furniture packaged in boxes. Packaging accomplishes many objectives, as shown in Figure V-5.

- Brand names

Forest products company opinions on the value of brand names vary. Neither do marketing experts agree. In one collection of articles on marketing ("Marketing 90/91," 1990, Dushkin Publishing Group, Inc., Guilford, Connecticut) the first article states: "Branding products . . . no longer guarantees loyal customers. In one case after another, the old established brands have been supplanted . . . " Another article discusses several well known brands and reports

FIGURE V-5
Packaging is a Positioning Tool

Packaging helps to accomplish positioning objectives by:

Establishing product value: The look of the package conveys the product value. The assemble-it-yourself cabinet can give the impression of a flimsy piece in one kind of package, a sturdy piece in another. The industrial drum shaped package of the product Janitor in a Drum positions it as an effective industrial strength cleanser.

Widening appeal: To satisfy the needs of customers who are not building large construction projects, Georgia-Pacific advertises small package lots of lumber, following the practice of fruit juice producers who may get 35 percent to 50 percent more for single serving packages than for a larger can. A packaging change can reach new customer segments, as Campbell Soup Company found when it added V-8 Juice packaged in asceptic cartons to appeal to the yuppie market.

Narrowing appeal: Package colors can be changed to appeal to men (deep gray), or to women (brighter colors). The type of dispenser can enhance the appeal to a very narrow segment of the market. For example, Noxzema lotion gained sales by packaging in a pump dispenser for women with long fingernails who don't like to risk chipping their nails by opening jars.

Creating an image: A new package can create a new image. Welch Foods, Inc. halted a decline in grape products sales by creating a new, homey, all natural, high quality image with new packaging. The use of wood in gift boxes for food and beverages adds an image of quality to the gift—a product differential that can be used by the producer of the wood in sales of specialty boxes.

that these products show "just how precious long-lived brand names have become."

- Product use

Product use is a positioning tactic that wood product manufacturers could employ more often. Sundance Spas, of Chino, California illustrates how positioning and repositioning can make or break a product. Sundance outdoor whirlpool spas are typical of luxury items whose sales plummet in a recession. Prior to the 1982 recession, Sundance positioned its spas to appeal to macho males between the ages of 25 to 49 who were lured by ads featuring spas with comely blondes holding champagne glasses. When the macho males no longer flocked to the spa showrooms, Sundance began to pay attention to other customers, particularly families with kids. More and more, these families were choosing to use the money they would otherwise spend on vacations for year round entertainment at home: A spa provided the entire family with fun at home for 365 days a year, and improved the value of their home besides.

Sundance recognized a new market opportunity and quickly changed its advertising, product, promotion, and distribution to position its spas for the new market. New ads replaced the comely blondes with families in the spa. Spa stories were featured in home improvement sections of newspapers, and spas were enlarged to accommodate a family. Maintenance was also simplified, and colors were changed. An education program for dealers helped transfer their enthusiasm from the macho male to the family. The repositioning strategy of making spas a 365-day-a-year vacation increased sales.

- Product quality

Positioning hardwood dimension lumber and components for sale to the furniture industry emphasizes product quality. Surveys have shown that industrial buyers of dimension lumber are concerned about quality control (assurance that each piece supplied will meet the buyer's standards), accuracy (the physical characteristics and allowable deviation, such as length, thickness, width, and weight), and delivery time. Dimension manufacturers use the advantage of quality to position their products with respect to their competitor's as dependable, reliable suppliers of quality dimension lumber.

- Promotion

Advertising, public relations, point of purchase (POP) displays, and special events communicate the manufacturer's positioning intent.

Advertising is often the first communication the potential buyer receives from the manufacturer.

Sears used promotion activities to reposition its products. Sears instituted a program to broaden its appeal to the market segment of younger, style-minded buyers while keeping traditional Sears customers—older, middle class and blue collar families. To accomplish the repositioning Sears used:

- ❑ Clothing lines with celebrity signatures
- ❑ Fashion boutiques in its stores
- ❑ A new line of children's clothing, called McKids (using the lure of the familiar McDonalds)
- ❑ New, smaller stores
- ❑ Specialty catalogs
- ❑ Merchandise tailored to regional tastes

Positioning by Providing Service

Customer service has been called the battleground of modern marketing. Convenience is one of the powerful weapons in the battle for positioning in modern society—convenience products, convenience retailing, convenience services, anything that saves time. Families that rely on the income of both spouses to maintain a lifestyle that includes on average two television sets, six radios, a VCR, and a stereo, look for products that save time and stores that are convenient to shop in. Convenience is prized by both industrial customers and consumers.

The demand for convenience is strikingly evident in food products. Consumers want food that is easy to prepare, fresher, portion controlled, requiring little or no clean up, good tasting, and good for them. They look for stores that are convenient to shop in and food that goes from shopping bag to the table in minimum time. Food product manufacturers are hard pressed to respond as quickly as the food marketplace changes.

Georgia-Pacific uses convenience as a positioning tool. GP features convenient package sizes in its advertisements to home center stores, and convenience is a product benefit frequently cited.

Positioning by offering convenience is a marketing tool increasingly utilized by business.

The growth of Duds'n Suds, an Iowa-based self-service laundromat chain, demonstrates how positioning through service can boost a company's sales. Duds'n Suds revolutionized the self service laundry industry by meeting

> basic needs of laundromat users. Every word in its slogan "good clean fun" became a positioning statement. "Good" referred to the quality service desired by families, who were targeted as a primary customers. "Clean" emphasized the home-like environment. "Fun" described the experiential satisfaction of emotional, physical, or mental stimulation.
>
> Good clean fun is provided by a home-like arrangement with couches and a large screen TV. The "family room" is complete with billiard table and video game machines. The "dining room" has tables and chairs where patrons can sit and enjoy popcorn, cookies, soft drinks, and beer from the snack bar. Beer, the original carrot to attract midwestern college students to the new laundromat concept, is no longer emphasized as a positioning device. More families are attracted by the bright, sparkling clean, conveniently arranged washers and dryers, snacks, and other amenities that turn the drudgery of the weekly wash to a family outing.

Extrapolate this concept to marketing wood for craftsmen. Provide an opportunity to work that wood in a clean, bright facility, equipped with woodworking machinery, with a large screen TV in front of some couches, a VCR for instructional tapes, and a snack bar to entertain the family while mom or dad puts together a bookcase.

Positioning Products For Two Special Market Segments

Two market segments, one on each end of the demographic scale, offer special opportunities for positioning: the children's market and the mature market.

The Children's Market

In 1988, the United States had 30 million children between the ages of four and 12. These children directly stimulated purchases of $6 to $8 billion in toys and games. These 30 million children really form three markets: a primary market of children who directly spend over $4 billion on products; a secondary market that influences parents and grandparents to spend about $50 billion on clothes, fast foods, breakfast foods, soft drinks, and snacks for children; and a future market of children whose tastes and brand preferences are now being molded.

Retailers have discovered children as a primary market, but the manufacturers of children's products pay relatively little attention to the direct wants and needs of children as buyers. They concentrate on the parents and grandparents.

Wood products manufacturers have neglected to address children directly. The toy and game industries consume lumber in making wooden toys or toys that use some wood. Toys and other products bought or used by children can use standard positioning concepts, if they are specially adapted for children.

Children's toys and furnishings made of wood are enduring and popular. Wood has lost market share to other materials, but part of that loss may be attributed to poor positioning. The market is large enough to deserve more attention by wood products manufacturers. Some of the points that may be helpful to producers of toys and other items for children:

❑ Children hold a different view of product benefits. Price means little, except perhaps for older children. Their priorities do not coincide with those of their parents. Children respond to special product features—if the product feels good, tastes good, smells good, fills their need for play, or for belonging. Gatorade is positioned as benefiting adults because it replenishes body fluids, but Gatorade appeals to children because it tastes good. Wood blocks appeal to a stability and feel that children like.

❑ Utilitarian products appeal to children, if they relate to their needs. Children could be targeted as customers for batteries, because they are essential to operate many electronic toys and games.

❑ Positioning techniques directed to children can be effective motivators:

❑ Children like bright colors. Package items in these colors.

❑ Brand names are powerful tools for children. They express brand preferences as early as age four.

❑ Product quality is important to children. If a toy breaks on first usage or with little abuse, children feel "ripped off" much as their parents would.

❑ The influence of television on children's purchasing behavior is evident.

❑ Point of Purchase (POP) displays directed at children can have a major impact.

❑ Positioning through special events is effective with children. McDonald's appeals to children by recognizing their birthdays and providing birthday party services. Birthday parties and playgrounds appeal both to children and their parents.

The Mature Market

On the other end of the buying spectrum, people now over 50 control over 75 percent of the nation's wealth and half of the discretionary income, according to 1989 estimates. People over 50 now account for only about 26 percent of the U.S. population, says the powerful American Association of Retired Persons (AARP), but they make 80 percent of the purchases of leisure travel, buy 50 percent of the domestic cars and silverware, and account for 44 percent of the purchases for home remodeling. In another ten years, when the first baby boomers born after World War II start hitting 50, these percentages will skyrocket.

Positioning products to reach this market will bring challenges marketers never dreamed of. The new 50-plus consumers are wealthier, healthier, better educated, more aware of fitness, more discriminating, and more sophisticated in many ways than their parents or grandparents. Manufacturers have not yet come to grips with the mature market and the missteps in advertising and positioning products for the over 50 are becoming legendary. Some companies have attempted, and failed, to reach this market segment by merely putting a new label on an existing product.

> Gerber Products Co., for years securely positioned in the baby market, tried to appeal to older customers with a product called, Singles. Singles never took off, partly because the baby food image embarrassed older consumers. Advertisers have also found that any suggestion that a product is for old people is doomed, just as they would not suggest that any product destined for people in their early 20s is for kids.

Persons over 50 are as difficult to classify as those under 30. There is a world of difference between a seven year old, a 17 year old, and a 27 year old. Similarly, the differences between older people (55 to 64) and very old people,(85 and older) spans a generation. In between are those classified as elderly (65 to 74) and aged (75 to 84).

These age classifications indicate that the mature market is just as segmented, as any other market—demographically, socioeconomically, geographically, psychographically, by health status, and behavior patterns.

Marketing to older Americans, as in marketing to any segment is based on meeting their needs and wants. Position products to appeal to mature persons as to any other age segment, focusing on the motivations and stresses that influence their purchases. The real key to positioning products and services for the mature segment is to recognize that their priorities differ from younger segments.

- ❑ Comfort becomes more important as people age. Honda auto advertising stresses room and comfort for mature buyers, instead of speed and performance.
- ❑ Physical security means more. Home security systems, larger, safer automobiles, and financial security are high on the list.
- ❑ As people mature, they prize convenience and products that help them accomplish tasks that grow more difficult with age, rather than the time savers that appeal to the younger buyers—remote controlled appliances, motorized carts, preprogrammed keys for microwaves ovens, battery operated pill boxes with reminders about taking the pills, and home deliveries are some examples.
- ❑ As nests empty and people retire, sociability becomes a necessity. Mature people want to feel useful. Products that foster personal growth—travel, lifelong learning, and interaction with younger people are priorities for many.

- Interests differ between age subsegments: Between the ages of 40 and 60 interest often shifts from acquiring possessions to "catered experiences" (visiting restaurants, art shows, or sporting events). From 60 to 80 the focus moves to "being experiences" (relationships with other people, philosophical interests, or religious experiences). Over the age of 80 the interests include simpler things, like taking a walk and engaging in friendly conversation.

Within these broad age generalizations are a highly diverse group of people with different lifestyles and values that provide clues to positioning products for older Americans.

The National Association for Senior Living Industries (NASLI) studied lifestyles and values of older adults and identified six psychographic segments that provide clues for positioning:

- Martyrs, representing 26 percent of the population over 55, are the largest group. Martyrs resist change. They feel that children should help parents, and often live with their children or relatives. Martyrs are next to the least educated, the most self denying, the least wealthy and the least healthy.

- Explorers, at 22 percent, are the next largest segment. They tend to be self reliant, most introverted, and slightly younger (65). In selecting a retirement home they are not as likely to believe cultural services are necessary.

- Pragmatists, at 21 percent, follow hard on the heels of explorers. Quite extroverted, quite and self indulgent, they are less educated and wealthy, conservative, and conformist. They are not interested in living in adult communities but like the comfort of security guards.

- Adapters, with an average age of 74, represent 11 percent of the older group, and are the most extroverted and open to change. Educated, self indulgent, healthy, and relatively wealthy, they appreciate conveniences and material possessions. They are interested in moving to a better climate or changing to a new lifestyle.

- Preservers, whose average age is a little older than adapters, constitute another 11 percent. The least healthy, they are need-driven and depend on helpers. Security is important to them. They tend to live with their children or rent an apartment in older adult housing in a high rise building. They are a good market for safety equipment.

- Attainers are the smallest segment. With a median age of 60, they are most autonomous, self indulgent, healthy, wealthy, best educated. They are open to change and least willing to believe children have obligations to help their parents. They own homes and are least likely to want to move to retirement community. They are more likely to want communities with children, recreational amenities, and more and larger bathrooms.

The psychographic segment to which a person belongs changes with the years. The attainer at 60 years of age may be driven into the preserver segment on reaching 75 as health problems develop. A mail order catalog, such as "Comfortably Yours," that offers safety items such as non-slip bathmats might appeal to preservers but have no interest for the younger explorers or attainers. The mature market offers great opportunities to manufacturers of factory made housing, furniture, and any wood items that enhance convenience and security. The manufacturer loses the opportunity, however, by failing to design and position products to meet the special needs and wants of each segment.

VI

TO MARKET, TO MARKET

. . . moving the product from the factory to the customer

CHAPTER SUMMARY *"The stores don't have it." complains a consumer. Industrial purchasers echo that complaint when required components are not readily available. Distribution, the science of making products available when and where customers are ready to purchase them, is the final link in the marketing process. Without a suitable distribution system, product design and differentiation, positioning to attract targeted customers, and product promotion are a futile exercise.*

Companies may make that vital contact with the customer directly or indirectly through many types of middlemen who provide the services that join manufacturer and customer. In marketing terms, there are many channels, or paths, of distribution. The choice of distribution channel is critical to a company's success.

Wood products manufacturers have never before had so many distribution channels from which to choose. The choice often determines the competitive advantage of a product. The newly developed channels make it easier to reach the customer with new products or with old products in new forms.

Distribution (called *place* in the marketer's vocabulary) involves the physical movement and transfer of ownership of a product from the producer to the customer. Distribution is carried out through marketing channels, the paths products follow on the way to the customer.

A manufacturer may design a product that fills wants and needs, carefully target customers for the product, develop differential benefits, and position the products so that those customers look for the product. All these objectives may be accomplished and still the product will fail if it is not moved out of the factory and into the hands of the customer. Boxes of merchandise stacked to a manufacturer's ceiling spell inventory, not opportunity.

> A long established book publishing company was acquired by a large conglomerate. The new management promptly installed a new computerized system to handle the business of selling the books. The ledgers of the merged company showed zooming sales and hefty accounts receivables.

> But complaints began to come in from book dealers. The books they ordered failed to arrive. Strange, because records showed an ample inventory of books. The company went through two vice presidents whose assignments were to straighten out the system. Finally, management discovered what the stockroom clerks knew all along. Somehow, the new system failed to provide for moving the books out of inventory to the shelves of book dealers.

Too often, distribution is the neglected side of marketing. Distribution decisions are made after product development and promotion. As noted by marketing experts, Louis D. Stern and Frederick D. Sturdivant, in the *Harvard Business Review* July/August 1987, "although American companies have been ignoring the ways in which they deliver products and services, their customers are increasingly inclined to demand higher standards of performance. And so, important opportunities for gaining a competitive advantage through distribution remain . . . ."

Distribution is the least flexible variable in marketing. The distribution decision involves long term commitments to middlemen or retailers. The structure of an industry often forces a firm to make distribution arrangements that may not be the best way to distribute the product. As a matter of practice, most marketers are limited in the distribution channels available to them. Many channel choices require large investments. Radio Shack, for example, markets its audio and electronic products through company owned stores. General Motors markets its autos through franchised dealers. The selection of an inappropriate channel, such as an unaggressive wholesaler, may also severely limit sales.

Some companies are more concerned with moving the product off the shipping platform into the first link in the distribution chain than with satisfying customer needs. The manufacturer who focuses on the last link in the distribution chain, the ultimate users, knows who the customers are, where they are, and their purchasing habits.

Marketing channels perform three functions, each one adding to the utility of the product for the customer:

- ❑ Time functions—having products and services for sale when the customer wants to purchase them
- ❑ Place functions—making products and services available in a location convenient for the customer
- ❑ Ownership functions—arranging for title to the product to pass from the manufacturer or his intermediary to the purchaser.

Toy manufacturers exhibit their new offerings at the annual toy fair in New York City each February to make the products available for sale when the toy buyers for department stores, toy shops, variety stores, and discount stores want them. Buyers make their decisions early to assure that the toys are produced and transported to their stores for the next Christmas season.

Retail stores provide a convenient place for buying transactions. Mail order catalogs and vending machines also provide a place for the customer to buy the product. At the retail store the consumer also obtains physical possession of and title to the product.

DISTRIBUTION CHANNELS—MEETING CUSTOMER NEEDS

Manufacturers look for a distribution method that best meets their customers' needs. The major distribution decision is whether to market directly to customers, or indirectly through a middleman, a business firm that stands between the producer and the customer.

Manufacturers of hardwood dimension components and other industrial wood products more frequently market direct to the customer. Manufacturers of consumer housewares usually use middlemen.

The middleman may be a wholesaler (also called a jobber or a distributor), or an agent, or broker. Distribution of some consumer products, like candy, usually goes from agent or broker through a wholesaler to a retailer on its way to the consumer.

Thousands of small manufacturers that produce a limited number of products use wholesalers to reach the hundreds of retailers who stock the products. Retailers count on wholesalers to provide a variety of products for their inventory. Wholesalers effectively act as the sales department of manufacturers they represent. Wholesalers buy lumber from the sawmills and may sell directly to lumber yards, to builder dealers, or drop ship.

Physical Distribution

Physical distribution—the logistics of physically transporting, storing, handling, inventorying, communicating, and data processing-moves the product where it is wanted at the time it is wanted. Physical distribution demands specialized experience and knowledge outside the scope of this book. Decisions to locate at a particular site are often based on the cost and logistics of physical distribution, which affect the ultimate price to the consumer.

> Pride Manufacturing Company, of Guilford, Maine manufactures wholesale wooden components, such as axles and wheels for toys, wood balls and beads, and furniture parts. Pride's decision to build a plant in the town of Florence, Wisconsin was strongly influenced by the town's proximity to a roadway network that allowed high truck load limits, thereby reducing the

cost of freight. The availability of white birch at a competitive price and factors that made the town a good place to do business were other basic factors in Pride's decision.

DIFFERENCES IN DISTRIBUTING INDUSTRIAL AND CONSUMER WOOD PRODUCTS

Distribution of industrial products differs in four ways from distribution of consumer products:

- Number of channels

  The industrial market uses fewer channels to reach customers. About three quarters of all industrial products are estimated to be sold directly to customers. The remainder are sold through agents or brokers directly to the industrial user, or from the agent through the wholesaler to the consumer.

  Less than 5 percent of all consumer products traditionally move directly from the producer to customers, but the trend to direct distribution is growing. Selling to consumers may involve numerous middlemen. Producers may sell to brokers, who sell to wholesalers, who then sell to jobbers, who in turn sell to subjobbers, and possibly others before reaching the consumer.

- Types of middlemen

  Middlemen who handle industrial wood products include wholesalers, drop shippers, manufacturers' representatives, and agents or commission salesmen. Retailers are usually not employed.

  Consumers usually depend on retailers to make products available. Manufacturers of consumer products may sell directly to consumers through their own retail stores, or they may sell indirectly through wholesalers, who in turn sell to retailers, who then sell to the consumer. The growth of home center stores, such as Hechingers on the East Coast, provides a new channel of distribution of products that are sold directly to the consumer market.

- Choice of channels

  Manufacturers of industrial products have fewer choices in distribution channels than manufacturers of consumer products. Industrial manufacturers' choices range from distributors who sell directly to industry, or manufacturers representatives or brokers who sell to distributors, who in turn sell to industry.

  Manufacturers of consumer products are searching for new (non traditional) distribution channels to better position their products.

Pantyhose is traditionally sold in department stores and upscale outlets. When L'eggs Products, Inc. (owned by Sara Lee Corp.) put L'eggs pantyhose on the market, they packaged it in plastic eggs, and shifted distribution to grocery and drug stores and mass merchandisers. L'eggs pantyhose were marketed as a convenience item, rather than a shopping product, so convenience stores were the logical distribution channel.

Through company owned retail stores, mail order catalogs, and direct mail databases, manufacturers are developing new channels for distribution of consumer products. Cable TV sales channels that display products and prices and allow viewers to phone in purchase orders, using their credit cards for payment, have opened new opportunities for distributing consumer products.

The IKEA stores are a new kind of furniture store offering, under one roof, a selection of 14,000 low-priced furniture and home furnishing items. Prices are low because customers serve themselves and most furniture items are sold ready to assemble (knocked down). IKEA also offers mail order service through a nationally distributed catalog.

❑ Expectations of service

The distribution channel for industrial products handles fewer transactions and usually larger orders. Specification selling—listing detailed physical and performance requirements for the industrial product to be considered for purchase—is prevalent. Industrial products emphasize service, technical assistance, field warehousing, and field inventory control.

While service is often important in selling durable products to consumers, few channel members have been expected to provide technical assistance to consumers. This expectation is changing as home centers and lumber yards add service and technical assistance to attract consumers.

The distinctions between distribution of industrial and consumer products are blurring. Many industrial manufacturers already use mail order catalogs. Dedicated cable TV sales channels can be adapted to display industrial products and have purchasing agents phone in orders. The increasing cost of personal selling is the prime motivation for changing methods of distribution.

CHOOSING THE APPROPRIATE CHANNEL FOR A WOOD PRODUCT

A number of perplexing questions enter into selecting a distribution strategy. For example, should the company enter into an intensive distribution arrangement using as many outlets as available, give one distributor exclusive rights, or use several types of middlemen?

Should sales be made directly to customers, bypassing the middleman? If so, will mail order catalogs be cost effective? How about selling via TV programs direct mail?

If distributing through a middleman appears a feasible choice, which type: distributors, manufacturer's representatives, wholesalers, or brokers?

No one marketing channel is superior to others. The choice of channel for each product depends on many factors, such as:

- ❑ The customer segment at which the product is targeted
- ❑ The characteristics of the product
- ❑ The type of firm
- ❑ Distributor cooperation

The industrial customer segment usually calls for direct manufacturer-to-buyer sales. The manufacturer whose customers are clustered in a few regions would select direct distribution, especially if buyers place small numbers of large orders.

Consumer products usually require a middleman to move the product from factory to a retail distributor, who usually serves a local market. New trends have removed the middleman from the distribution chain of some products: the growth of self service, more sophisticated mail order houses, and purchasing from neighbors such as through Tupperware parties.

As a general rule, the more standardized the product, the more middlemen there are involved in distribution. Another general rule is that the lower the unit value of a product, the more middlemen there are involved. A manufacturing company whose product required specialized knowledge or technical know-how would find middlemen to meet customer service needs. Complex products, such as custom made industrial installations, are sold directly to the buyer.

Companies with financial, management, and marketing skills are less inclined to use middlemen. Financially strong firms may vertically integrate marketing to control all the functions from production through marketing to the ultimate consumer. Many of the major lumber producers in the United States control their timber resources, processing facilities, and distribution. Other companies have only forward integration, wherein they control processing and distribution. A company with a vertical marketing system controls design, manufacture, brand, price, promotion, purchasing, stocking, displaying, selling, delivery, and often financing.

Others are integrated backwards to control timber resources and the design and manufacture. Almost every post World War II recession has been accompanied by large producers buying smaller sawmills to gain ownership or access to timber.

Larger retailers, such as Safeway, control part of their manufacturing facilities, which produce private label products. The Holiday Inns owns a carpet mill and a furniture manufacturer to furnish its hotel chain.

Vertical integration isn't limited to large companies. Smaller firms can enjoy the benefits of controlling the entire system, from raw material to distribution, as well.

> Bergstrom Wood Products, Inc., of International Falls, Minnesota typifies the smaller, vertically integrated firm. The family owned company has access to stands of Aspen timber. Bergstrom crews log the timber, process the logs to lumber, transport it to their pallet manufacturing plant, and market custom pallets directly to a local paper mill.

A distributor offers some wood products firms a distinct advantage; others may be better off without one. Companies that benefit from a distributor's services are small firms that cannot justify supporting a sales staff and want to lessen credit risks with customers. Distributors provide credit and technical support in exchange for the company yielding some degree of contact with their customers.

A distributor also enables the manufacturer to concentrate on manufacturing rather than dissipating efforts on training a sales force. For small manufacturers, wholesalers can help ease cash flow problems because they are accustomed to paying for items upon shipment rather than when they sell to their customers. When there are many small orders purchased by many customers, accounting transactions become a high percentage of the cost of doing business. Wholesalers are set up to handle these transactions.

If a distributor fails to adequately promote a product, the manufacturer is often forced to develop a new or different way to distribute products. Avon cosmetics was forced into its famous direct selling through Avon ladies who called on their neighbors because cosmetics distributors failed to adequately promote Avon products. A manufacturer of custom made furniture claimed that the furniture line never achieved its sales potential until the company began to market directly to customers. Their assumption was confirmed when sales took off.

INNOVATIVE DISTRIBUTION

Today, distribution strategy focuses on reaching consumers where they shop, where they work, or where they live. Customer focused distribution is forcing the channels out of their straight, traditional paths. The quest to reach customers follows many channels, some that were not recognized until recently. Distribution lends itself to innovation, just as do products and promotion. Innovative distribution may be the way to catch up with changes in customers' buying habits, or open the door for a new manufacturer to get around an established, closed distribution system. New players and new distribution routes are opening new opportunities for wood products manufacturers. For example:

- ❑ Telephone expeditors eliminate a layer in the distribution process by buying in volume from manufacturers and selling directly to retailers. Prevalent in supplying auto parts stores and garages, the use of telephone expeditors is tempting other suppliers who are hoping to increase profits by bypassing traditional channels.

- ❑ Warehouse distributors meet the needs of specialty shops by offering sizeable discounts that compete with the traditional channels.

- ❑ The packaged goods concept changes channels with the changes in the market. For example, Fuji Photo Film U.S.A. Inc. used to sell its floppy disks for computers through traditional contract stationers with sales forces that sold to office supply stores. As the personal computer became a household item in 30 percent of United States homes, Fuji changed the distribution concept of floppy disks from specialty items to packaged goods, which are now sold through food and drug chains.

- ❑ Non traditional direct marketing reaches the customers where they live.

  When Tripledge windshield wipers, which are guaranteed to last the life of a buyer's car, tried to sell through retailers it ran into problems. When the company switched to a strategy designed to reach customers directly, sales took off. Customers were reached through credit card inserts, catalogs, print advertising, TV "infomercials," and print omnibus ads that feature several items.

- ❑ Servicing channels reach consumers through a channel that provides a service-like assembly and delivery with the product.

  Garden Way Inc. traditionally sold its garden tillers direct by mail to customers who then assembled the tillers when they received them at home. As the number of suburban customers increased, Garden Way realized that these consumers were ill equipped to handle the 300 pound box unloaded in their driveway, nor could they readily assemble the machine. Shifting to independent power equipment dealers whose sales forces were knowledgeable and equipped to assemble and service machines, Garden Way was able to expand into the suburban market.

Changing channels is achieving increased acceptance as a way to gain a competitive edge. However, it has its pitfalls, as explained in Figure VI-1.

FIGURE VI-1
Changing Distribution Channels

When sales fail to live up to expectations, changing the method or channel of distribution may be the solution.

Manufacturers of niche products sometimes find changing from selling through distributors to selling direct to consumers increases sales. Others have found that the additional costs of handling and processing individual orders from thousands of customers instead of one distributor decrease profits and cash flow.

If a company is planning to change distribution channels, this is what distribution experts advise:

- ❑ When adding another channel of distribution (for example, direct mail), provide incentives for the current distributor to continue handling the product.
- ❑ Determine as early as possible what the company will gain by changing channels. Is it the ability to provide superior service, or a better pricing structure? If the new channels don't produce more in sales or a favorable customer reaction, reconsider the change.
- ❑ Assess the problems and costs of meeting distribution requirements. Can the company handle daily orders? Will it be necessary to increase inventory? Can smaller orders be processed?
- ❑ Will the product require repackaging for the new channel? For example, distribution through a warehouse club requires an economy sized package that can be sold from a pallet. Distributing through some chains requires in-store merchandising.
- ❑ Check what competitors are doing. Does the company have the financial resources to compete head-on with a manufacturer who already occupies substantial shelf space in stores? If not, put time into designing innovative methods of distribution.

Adapted from "Tips on Changing Channels," in Sales & Marketing Management, August 1990.

MIDDLEMEN

Brokers, agents, wholesalers, distributors, and retailers each play a role in linking the producer and customer. Each functions under a set of business rules that determines how they are paid.

Wholesalers

Two types of wholesalers provide services: Merchant wholesalers and agent wholesalers, (also called manufacturer's representatives).

Merchant wholesalers are middlemen who take title to the products they handle. Taking title marks a major difference between merchant wholesalers and agents or brokers, who do not take title to products.

Merchant wholesalers provide support to small companies:

- For small, underfinanced producers and new entries into a market, eliminating or reducing the need for a costly sales force
- Purchasing and paying for products in advance of sales, improving cash flow
- Sharing the risk of marketing and distributing by warehousing products
- Reducing the cost of inventory, and increasing the responsiveness to market demands by reselling in smaller quantities required by customers
- Providing a valuable source of market information to producers
- Providing credit to customers to whom the individual producer might not be able to extend credit

For small retailers, full function merchant wholesalers are almost indispensable in stocking small quantities of many essential items.

Despite the services they perform for both producers and customers, merchant wholesalers are often maligned by producers, consumer organizations, retailers, and the media. Discount retailers advertise that they can offer lower prices by eliminating the middleman. Chain stores brag that their lower prices come from buying direct from the manufacturer. Direct manufacturers' outlets are the retailing rage of the '90s.

The role of the wholesaler in the forest products industry is subject to almost constant scrutiny. Some wholesalers assert that the trend to JIT deliveries have made the wholesaler more important than ever to the industry.

The August 17, 1990 *Random Lengths* newsletter, for North American forest products marketers, reported that a survey of wholesalers showed mixed opinions of themselves. Half of the respondents, especially those with distribution facilities, thought that wholesalers were more important than five years ago; 33 percent thought the situation unchanged, and 17 percent said they were less important. One wholesaler noted that "a lot of companies have found out that buying direct isn't all that glamorous. They need the representation and credit that wholesalers allow."

Wholesalers that provide full functions have high operating expenses, averaging from almost 14 percent to as high as 20 percent of sales. In an effort to reduce costs, several types of limited-function merchant wholesalers have evolved:

- Cash and Carry wholesalers perform most wholesaling functions, with the exception of delivery and financing. By eliminating those two services, cash and carry wholesalers report a reduction in overhead of 9 percent of sales.
- Truck wholesalers or truck jobbers make regular deliveries to retail stores, promote their product lines, sell and collect. Margins are low and delivery costs high, increasing overhead costs of these wholesalers to 15 percent.
- Drop shippers are familiar to lumber producers. These wholesalers can assemble various types and grades of lumber to fill a customer's order; they maintain no inventory of the lumber, eliminating the cost of loading and unloading carload shipments, and storage. They forward orders from customers to producers, who in turn ship directly to the customers. Drop shippers never physically handle the lumber, although they take title, and are therefore able to reduce their operating costs to 4 percent to 5 percent of sales.

Independent Wholesalers

In addition to wholesalers, manufacturers of wood products may select one of three types of independent wholesaling middlemen, who differ in several important respects, as shown in Figure VI-2.

- Manufacturers' representatives operate on a set commission rate, with commissions paid only on actual sales, removing much of the guesswork from the cost of selling. A manufacturer often works with several manufacturers' representatives, each covering a specific territory. These reps keep their commissions at a relatively low percent of sales by spreading their costs over a number of related but non-competing products. A rep specializing in housewares may also sell an assortment of wooden specialty items, such as cannisters, bread boxes, and bread boards. Commissions vary from about 5 percent for the lumber industry to close to 9 percent for the furniture industry, depending upon other services provided, such as warehousing and help in new product development.

FIGURE VI-2
Differences Between Types of Middlemen

Item	Wholesalers	Manufacturer's Reps and Broker
Title and Possession of Product	Take possession	Do not take possession
Compensation	Mark up cost from manufacturer	Fixed commission on products sold
Customer credit	Provide trade credit	Do not provide credit
Inventory	Pay for storage and delivery	Do not inventory or deliver
Sales function	Sales force	Match buyers with sellers
Promotion	Promote products	Usually do not promote

The manufacturers rep assists manufacturers in selling and distribution by:

- ❑ Eliminating the expense of a direct sales force that continues whether or not sales are made
- ❑ Minimizing the time and costs of training personnel
- ❑ Giving a manufacturer immediate access to the market
- ❑ Providing manufacturers greater marketing flexibility, an advantage for small and medium sized companies and those entering a new market
- ❑ Using knowledge of the industry, contacts, and coverage to expand sales
- ❑ Brokers act as one time intermediaries between buyers and sellers where the number of potential suppliers and purchasers is large, as in real estate and used machinery.

❑ Selling agents have played an important role in selling lumber and wood products produced by small, usually poorly financed firms. They serve as independent marketing departments, often handling the total marketing program for a firm's product line. Many small producers prefer selling agents because the agents relieve them of marketing activities, leaving the producer to concentrate on production. Operating expenses of selling agents have been reported at 3.2 percent of sales. The use of selling agents has declined as manufacturers have become more marketing oriented and interested in better control of their sales.

SPECIALIZED DISTRIBUTION CHANNELS

Mail Order Catalog Marketing

Direct mail order catalogs reach customers where they work and consumers where they live. Direct mail catalogs function as hard working sales representatives for industrial products and persuasive salesmen for consumer products.

Mail Order Catalog Marketing of Industrial Products

Over 1.1 billion business-to-business catalogs were mailed in 1990, according to Ed Burnett, president of Ed Burnett Consultants, Inc. of Englewood, N.J. Mail order catalog sales were reported at $50.4 billion in 1989. Thomas Register catalog company estimates that catalog orders through its service total about $400 million per day. A 1990 research report found that 98 percent of specifiers and buyers reported that they need catalogs from manufacturers in their work.

Catalogs have several advantages over other types of sales tools. For example, they provide readily available printed technical specifications and price quotes for ordering. They can also be filed as a permanent reminder, and they are available on customer's shelves 24 hours a day, seven days a week. Catalogs may cost millions of dollars or consist of simple copied, stapled technical bulletins, booklets, or brochures. Whatever their form or cost, the appearance of catalogs should reflect the quality of the items being sold.

Desktop publishing software has brought professional looking catalogs within the budget of smaller companies. Some suggestions from professionals for manufacturers who create their own catalogs are:

- ❑ Decide how much the company can afford to spend for a catalog
- ❑ Determine how many catalogs to print and select a binding that fits the catalog's budget and purpose
- ❑ Determine the page size and paper stock
- ❑ Decide what items the catalog will feature
- ❑ Group the products by type and color code to make them easier to locate
- ❑ Provide the information purchasers require to specify the products
- ❑ Use photos, drawings, and illustrations as much as possible

- ❑ Provide comprehensive information and prices for each item
- ❑ Provide an index and a table of contents
- ❑ Provide a simple order form, 800 numbers and fax numbers

The high tech revolution has introduced the paperless catalog—catalogs on disks, video taped brochures, and on line computer ordering. These high tech techniques fit well with JIT delivery and the concept of "stockless inventories." According to a study by the Wharton School of Business in Philadelphia, video brochures pay off in customer retention, which exceeds remembrance of print advertising by more than 50 percent. Moreover, prospective customers who view video tapes are reported to reach a positive buying decision 72 percent of the time.

Mail Order Catalog Marketing of Consumer Products

About 10,000 consumer mail order catalogs are published today, more than twice as many as in the 1960s. More than 12 billion catalogs are mailed annually, up from 4.7 billion in 1980. Today's mail order catalog is a far cry from the Sears general merchandise catalog that graced American living room tables in bygone eras. Many of these are imitations of successful catalogs, such as Land's End, which sells sportswear, or the high tech Sharper Image. The proliferation of catalogs has overwhelmed buyers. However, demographics still favor increased catalog sales. More women are working outside the home and are pressed for time to shop. Moreover, as the population ages, many persons—especially the frail elderly—are expected to prefer shopping from the comfort of their armchairs instead of fighting crowds in the mall and coping with traffic jams.

Most catalogs are aimed at niche markets, identified and tracked via computer data bases.

> Portland-based Norm Thompson, one of the largest mail order houses in the country, demonstrates the effectiveness of carefully targeted mail order. Mail orders account for 85 percent of Norm Thompson's estimated $50 million in annual sales. About 40 percent of these sales are made during the Christmas season. The Norm Thompson slogan, "escape from the ordinary" appeals to affluent, educated persons who are interested in quality products. With a "you be the judge" guarantee, Norm Thompson has positioned itself as a dependable quality company for quality people. To combat the catalog overload which has turned off some consumers, Norm Thompson is offering its products catalogs to an ever more selective list.

Direct mail catalogs are a useful distribution channel for small wood products and also for large products like furniture. One issue of the Damark International, Inc. of Minneapolis, Minnesota catalog includes a cedar dog house that can be "assembled in no time," priced from $79 for the medium sized to $129 for the extra large. A computer furniture package of "genuine oak finish with hardwood solids," a picnic parquet table with four benches made of "genuine California redwood," solid teak storage cases, a filing cabinet made of "solid oak with fine oak veneers," "genuine Ramin wood mini blinds with a 'rich pecan finish'," and a "solid oak " file trolley are among the offerings of the December 1988 catalog. Many of these items are brand named sold through the catalog at discount prices.

Figure VI-3 illustrates a variety of wood products distributed through direct mail catalogs.

Eddie Bauer's 1988 gift catalog offers "handsome aromatic cedar hangers," cedar blocks, cedar duck- and tree-shaped pieces to protect against insects, wood shoe trees, a rosewood log crib, a gift basket of firestarters molded of pitchwood from Douglas fir and ponderosa pine, and an assortment of firewood cribs made of wood. These items are finely crafted products sold at quality prices to appeal to buyers who appreciate quality. Some of these quality items are shown in Figure VI-4.

Mail order catalogs for handymen feature wood parts, such as plugs and buttons, dowels, balls, pegs, candlesticks, candle cups, novelty and specialty items, and tools and accessories for woodworking.

> Yield House of Conway, New Hampshire uses both a handsome mail order catalog and retail stores in the coastal Northeastern states to distribute furnishings of real wood. Yield House positions its furnishings to appeal to persons who appreciate quality and family values. The catalog states: "While it may cost a little more initially than particle board or wood substitutes, Yield House solid wood furniture represents an investment in quality as well as good looks. It's built to last—to be handed down from generation to generation." A Yield House catalog whets the appetite for treasured crafted pieces of wood furniture. Weight poses a minor barrier when marketing quality.

Reaching customers where they eat is another technique for direct mail.

> Stash Tea Company uses the restaurant table to tell customers about its catalog. When a restaurant customer asks for a cup of tea, the popular Stash Teas are served in individual foil pouches. Stash Tea uses the back of those pouches to introduce its catalog of teas. Printed on the back of each foil

FIGURE VI-3
A Variety of Wood Products are Distributed Through Direct Mail Catalogs

FIGURE VI-4
Quality Wood Items Shown in a Catalog

packet is an invitation to send for the catalog. Customers have eagerly accepted the invitation: 100,000 catalogs have been sent to consumers and 20 percent of them have made purchases averaging $25 each.

Technical Requirements for Mail Order

A mail order distribution channel demands rigorous attention to servicing the order. Back-end services, as direct marketers call them, cover everything required to fill an order once a customer makes a request. Typically, back-end services include the 10 steps shown in Figure VI-5.

Franchising

Franchising is professionally managed and centrally programmed to achieve economies of scale by virtue of size and elimination of duplicated services. Franchisees contract to meet operating requirements of the manufacturer or franchisor in exchange for being able to buy supplies at lower prices and receiving marketing, management, technical and financial services for a specified fee. The restaurant business, most notably McDonalds, Kentucky Fried Chicken, Burger King and other fast food restaurants, have embraced franchising. The top fast food franchises have proven themselves in tough international markets.

Franchises are not limited to restaurants. Energy conservation centers, gourmet cheese stores, and furniture stripping are franchised. Franchises have gained popularity by being sensitive to customer needs.

Hand tools are usually distributed by a two step process. Snap-On Tools broke the mold by selling more expensive and higher quality tools through franchised routes operated by independent drivers. Customers are willing to pay higher prices for the quality of the tools and the convenience of buying at the work site and on credit. California Closet Company manufactures and franchises a closet organizer custom designed to fit into the customer's closet to provide more clothes storage space than the ordinary closet. Distributed through a network of independently owned and operated franchises, California Closet has positioned itself to attract homeowners who are willing to invest around $300 to make their closets more efficient. Closets are a growing portion of the home remodeling business.

California Closet provides design services and installation by largely young carpenters who are specially trained. Franchise distribution is a natural for closet remodeling. The franchisees, who go by the name Creative Closets, Inc. outside of California, pay $45,000 for vans, equipment, and inventory. Five percent of the franchise revenue goes to the company. Another 5 percent is, by agreement, spent on local advertis-

FIGURE VI-5
Ten Steps in the Back End of Mail Order Distribution

1. The process begins with a good order form and clear instructions.
2. The less fuss to place an order, the more likely the customer is to place one. Decisions on receipt of orders—telephone, fax, mail, using an 800 number (hire a service company and use their 800 number or establish a company 800 number), and honoring credit cards are essential to the successful operation of a mail order business.
3. Credit approval is more critical in mail order, where the customer and his credit card are not visible to the order taker, than in the retail busness. Careful credit screening is essential.
4. Computer processing to track mailing lists, inventory, billing, and other processing requirements is available from specialized computer software companies.
5. Customer and prospect mailing lists are maintained by a computer software program. The tightly segmented mailing list is one of the keys to company survival as the number of mail order catalogs has proliferated.
6. Billing is another essential element of a mail order operation that can be handled with computer software.
7. Controlling inventory is essential to comply with legal delivery requirements and to minimize customer complaints. Slow deliveries or non-deliveries are among the most frequent consumer complaints.
8. Report preparation is facilitated by computer software programs, a necessity in an operation with hundreds of products and thousands of customers.
9. Filling and shipping orders is the process where the order forms, receipt procedures, credit approval, and inventory control come together. Maximum speed and accuracy in filling orders is reflected in the bottom line of a mail order operation.
10. Inevitably customer service is required in mail order. Shipping errors, exchanges, returns, complaints require handling. The way these are handled often determines whether a customer is retained or lost.

ing. California Closets carved out a $6 million a year niche in the home remodeling industry. President Neil Bolter believes that advertising has been crucial to the company's success. "We literally have to go out and educate the public," he says. "They've never heard about redecorating a closet before." Backing up that belief, California Closets spends about $600,000 of its earnings in advertising.

Franchising is a type of vertical marketing system (VMS) that coordinates distribution through contractual agreements. Contractual systems account for approximately 40 percent of retail sales. In addition to franchising, vertical marketing contractual agreements include:

- Wholesaler sponsored voluntary chains such as the 5,000 IGA stores and Western Auto Stores in which a group of retailers formally agrees to purchase a wholesaler's products, standardize facilities, and use a common name.
- Retail cooperatives in which a group of retailers set up a wholesaling operation in order to compete with the large chains. These cooperatives have been successful in the grocery industry, accounting for about 20 percent of grocery sales.

A second type of VMS is the corporate system where a single owner controls each stage of the marketing channel. It is reported that Sears purchases half its product lines from companies in which it holds an equity interest.

A third type is the administered VMS. In this VMS, the distribution is dominated by one powerful member who acts as the channel captain, or the controlling member of the distribution chain. Magnavox distributes through an administered VMS, obtaining strong support from independently owned and operated retailers and benefiting from the strength of its brand reputation.

Retailing

Retailers have been called "the last three feet of the marketing channel." Retailers come in all sizes from mom and pop stores to power retailers, located in a cubbyhole on Main Street or in a mass merchandise store spanning several city blocks, from exclusive specialty stores to discount houses. Teenagers and their grandparents go to the mall for recreation. Busy working mothers stop in the supermarket to buy dinner. Today's retailers are a far cry from the old country store, and yet, in many ways, the new forms of retailing and supermalls are filling many of the same functions as the country store described in the "Golden Book of America," by Irwin Shapiro.

> "The old country store was more than a store. It was an old men's clubhouse, a meeting place for friends and neighbors, a bank, a safe place where Pa could leave the children when he went to see a blacksmith about shoes for his horses. It was also a United States Post Office, fourth class."

The changing face of retailing has moved this channel of distribution into a position of more importance than ever to wood products manufacturers. Retailers are characterized according to:

- The effort required of shoppers

  Convenience and neighborhood groceries minimize travel by providing convenient locations in which the customers can quickly go in and out. Furniture and appliances are usually sold through shopping stores that encourage shoppers to tarry. Specialty retailers such as Neiman-Marcus, Tiffany & Co., and similar stores provide incentives for customers to shop at a more leisurely pace.

- The services provided to customers

  From self service in supermarkets to full service in specialty stores and department stores, customers can choose the level of service they prefer. In between are a selection of services, such as mail order and discount retailers, and door-to-door sales.

- The products carried

  Single line or limited lines of products distinguish Toys "R" Us™, Wickes (furniture), and Handy Man home repair products retailers. The difference between limited line and general merchandise retailers is emphasized in a statement by the marketing Vice President of Toys "R" Us™: "Sears can show customers three types of footballs, but we can show them forty." Specialty stores stock part of a single line of products in great depth. Specialty stores have become a new opportunity in retailing wood products with the development of stores specializing in cabinets, or doors, or windows.

  Window Headquarters, located in Brooklyn, New York specializes in windows for new and remodeled homes, including bay windows, patio doors, and custom windows. A customer planning to install windows knows that Window Headquarters has an exceptional selection. If you can't get it there, you can't get it. The seven-year-old company has about 30 employees and four branch offices that produce sales in excess of $6.6 million annually. Window Headquarters distributes both through retail and wholesale channels.

- Department stores

  Department stores carry many product lines and offer a wide variety of services from clothing and accessories, household linens and dry goods, home furnishings and furniture to beauty parlors and restaurants. Originally, large department stores were associated with particular cities—Macy's in New York, Harrod's in London, Au Printemps in Paris. To compete with discount stores and giant mass merchandisers (hypermarkets), department stores have added bargain basements, enlarged parking lots, opened branches in outlying suburban shopping centers, and even expanded into mail order catalog sales.

- ❑ Where the transaction takes place

  House-to-house retailing (Fuller Brush Company and World Book Encyclopedia) provides maximum convenience for the customer at home, as products are ordered from catalogs and delivered to customers' homes.

- ❑ Ownership

  Chain stores, many stores under a single ownership, provide economies of scale, while independent retailers provide flexibility and knowledge of the neighborhood.

An innovation in retailing are planned shopping centers, in which tenants are aggregated with care to provide a set of stores that complement each other. Similarly, neighborhood shopping centers provide convenient shopping for 5,000 to 15,000 shoppers. Community shopping centers serve 20,000 to 100,000 persons in a wider trade area, and regional shopping centers are located in areas where at least 250,000 people are within 30 minutes of the center.

Planned shopping centers account for about 40 percent of all retail sales in the United States.

Scrambled merchandising contrasts with a specialty store whose focus is narrow by expanding into dissimilar lines to generate additional sales volume. A modern drugstore is an example of a scrambled shopping center, with a mix of gift items, groceries, housewares, magazines, and small appliances in addition to prescription and over-the-counter drugs.

The future of retailing is summed up with one word: Change! Just as manufacturers are digging for niche products, retailers are searching for specialized merchandising styles. The changes in retailing reinforce the premise that distribution should be considered when the product is conceived, well before it appears as inventory on the manufacturer's shelf.

Nowhere is the change in retailing better illustrated than in marketing furniture. Multiple channels in furniture distribution offers manufacturers significant opportunities to match the type of furniture with distribution strategy.

Furniture manufacturers may select from any of at least 10 distribution options:

- ❑ Single unit full service furniture stores, such as specialty furniture stores

- ❑ Full service furniture chains
- ❑ Department stores—conventional, mass merchandise chains and discount department stores
- ❑ Warehouse showrooms
- ❑ Manufacturer owned stores, such as Ethan Allen Furniture
- ❑ Retail furniture stores
- ❑ Wholesale furniture distributors, who in turn sell through retail furniture stores or interior designers
- ❑ Mail order and telephone retailers
- ❑ Buying clubs, incentive houses, peddlers, discount furniture stores
- ❑ Video taped sales

The manufacturer of upscale RTA furniture would distribute through upscale home center stores and not through a discount store. The product image carries through to distribution to the customer.

REVERSE CHANNELS OR FULL CYCLE DISTRIBUTION

Distribution generally concerns the movement, or channeling, of products from the producer to the consumer or industrial user. The environmental movement has spurred an interest in moving or channeling used products from the consumer back to the producer. Reverse channels make possible the reuse or recycling of material, which is the ultimate goal of reverse or backward channel movement. The forest products industry is paying more attention to reverse channeling as timber becomes scarce and more expensive and recycling becomes an environmental and social objective. Reverse channeling has been pushed by the passage of additional legislation to control litter or require recycling. Recycling of soft drink containers is well-known in the states that have passed deposit laws. Retailers and bottlers in the soft drink industry provide the mechanism for the reverse channel.

Recycling is a growing trend in the forest products industry. Substantial investments by the pulp and paper industry in recycling facilities require active participation by consumers and distributors, in reverse channeling.

> After successfully operating 16 paper recycling plants all over the nation, in 1990 Weyerhaeuser Co. invested $5 million for equipment and construction of its 17th plant. Processing 24,000 tons a month, plant No. 17 is Weyerhaeuser's largest paper recycling center. Weyerhaeuser construction

> activity includes a $70 million cardboard recycling plant with a daily 400 ton capacity, to compensate for the decline in the timber supply. The cardboard processed by the facility equals the daily box waste from 250 large grocery stores, according to a company spokesman. The processed pulp is used to make heavy brown wrapping paper.

For reverse distribution channels to achieve their goals, certain conditions must be met. Assuming that a technology is available to efficiently use the recycled material:

- ❑ Consumers should have some incentive to make the effort to collect material for recycling. For many years, the Boy Scouts, churches, and charitable institutions have had projects to collect newspapers for recycling to earn funds for their organizations.
- ❑ A market should be continuously available for the recycled products. When recycling plants are over supplied and will not buy additional newspapers, consumers and suppliers tend to lose the recycling behavior habit.
- ❑ Households should be sufficiently motivated to expend the effort for recycling. Environmental concerns prompt consumers to bundle newspapers and transport them to collection sites.
- ❑ There should be an advantage to consumers to recycle rather than reuse products for their own purposes. In periods of rising energy prices, competitive uses for newspapers as fireplace fuel may discourage consumers from giving away their newspapers.

Manufacturers and businesses practice reverse distribution in their search for methods to reduce the cost of sending waste wood to landfills. Discarded and broken industrial pallets constitute a large source of wood waste.

> An Oregon company called Wood Exchange, Inc. uses reverse distribution to reduce the wood waste buried in Oregon landfills by up to 80,000 tons a year. Recapturing the wood fiber obtained from recycled clean waste wood from wood pallets, crating material, and construction debris, Oregon wood converts waste that would otherwise choke landfills into a raw material for useful products.
>
> The bulk of the material recycled at Wood Exchange comes from wood pallets used in distribution and warehousing. Discarded pallets and other wood waste are transported to the recycling facility. Repairable pallets are fixed and sold. Heavily damaged pallets are ground into wood chips that go into the manufacture of hardboard panels.
>
> The recycling serves a double purpose: It reduces the volume of waste wood going to landfills and it supplies wood fiber for manufacturing hardboard. When log shortages and a poor economy cause sawmills and ply-

> wood mills to shut down, a shortage of wood fiber develops. The lack of fiber from the usual trimmings, shavings, and wood chips developed in processing logs could also close down the hardboard plant were it not for recycled pallets.

To encourage manufacturers to invest in utilizing recycled material, they must have a profitable use for a substantial and continuing volume of the products made from the recycled material. Recycled wood fiber, newspapers, recycled aluminum, and reclaimed steel have ready markets.

VII

PROMOTION

. . . it doesn't have to cost an arm and a leg

CHAPTER SUMMARY *Promotion is the marketing term for communication with potential customers about the products and services a company offers. Companies have high expectations from their expenditures on promotion, and usually, programs that support the marketing strategy fill these expectations. Still, the promotion budget is one of the first to be pared during business downturns. But promotion doesn't have to cost the proverbial arm and a leg. Many powerful methods of promotion cost relatively little.*

Alternative lower cost techniques for personal selling, advertising, public relations, and sales promotion are suggested. This chapter also includes ways of analyzing potential customers as a basis for designing a promotion strategy. It also discusses the seven stages in closing a sale, types of media, guidelines for creating advertisements that work, suggestions for effective publicity, and sales promotion tools.

High impact marketing and new merchandising techniques developed by home improvement stores are two of the alternative methods for promoting wood products that manufacturers should find useful.

To many people, marketing means promoting. Promoting, in turn, is popularly equated with advertising. The company with money to pay for advertising enjoys an indisputable advantage over competitors. But advertising isn't the only avenue to promotion. Smaller firms with slim bankrolls have achieved dramatic results in promoting products.

Promotion is the marketing word for communicating with customers about products and services. Methods for promotion communication include:

❑ Personal selling

Face to face communication between sales people and customers. Immediate feedback from customers makes personal selling an art form of the highest order. Training and motivating salesmen constantly engages the attention of sales managers. Personal selling is typical for industrial products and high value products.

- ❑ Advertising

  Indirect communication from manufacturers to potential customers. In advertising, as in other forms of non-personal selling, the only method for gaging the effectiveness of an advertisement is by customer reaction, measured by product sales. Consumer products are usually advertised in mass media, such as newspapers, television, radio, billboards and magazines. Industrial products manufacturers advertise in trade magazines directed to industrial buyers. A company that places considerable advertising uses an advertising agency—a firm of professionals who help develop an advertising plan. Traditionally, the agency is compensated by a commission on advertising expenditures.

- ❑ Public relations

  Marketers say: "Advertising pays, but publicity helps it pay off." Public relations is communication with a company's various "publics" (also called "stakeholders")—customers, suppliers, stockholders, employees, the United States Congress, state and local governments, the general public, the media, and activist organizations of all types whose opinions have the power to influence the products companies make, how they make them, and how they sell them. Product *publicity* is public relations aimed at communicating with potential customers about a company's products and services. Compared to other forms of promotion, publicity costs are usually fairly low.

  Public relations departments of larger companies frequently employ a public affairs director who might lobby elected officials to protect the company's interests in matters of legislation.

- ❑ Sales promotion

  Communication activities that help customers understand the value of the product. Sales promotion includes displays, trade shows, expositions, demonstrations, POP tags, specialty advertising (useful articles that carry the advertiser's name), samples, coupons, premiums, contests, trading stamps, and similar methods.

EXPECTATIONS FROM PROMOTION

Promotion is often underbudgeted because it is the easiest marketing variable to skimp on when funds are tight. Although expectations from funds spent on promotion are high, advertising and sales promotion expenditures are typically among the first to be pared in cost cutting efforts.

Promotion is expected to:

- Stabilize sales
- Stimulate demand for products or services
- Communicate information about the product and its availability to potential customers
- Inform potential customers about the differential advantages of the product
- Enhance the value of a product to the potential customer by emphasizing status gratification or satisfying other aspects of consumer behavior
- Reduce fluctuations in demand and insulate the market position
- Build name recognition and create brand identity
- Build a positive company image

The Hardwood Manufacturer's Association (HMA) Promotion and Marketing Program uses a mix of public relations and sales promotion activities to promote solid wood products, including:

- Publication and distribution of a national consumer magazine, *Hardwood Expressions,* to more than 20,000 consumers, wood products manufacturers, and retailers.
- Development and distribution of syndicated stories for television and radio to inform consumers about the benefits of using hardwoods in the home.
- Creation of Harvey Hardwood, a solid red oak "spokestree" with an "oakly" voice that tells the hardwood story at malls. A 1990 tour, sponsored by *Good Housekeeping* magazine visited more than 25 malls.
- An internal newsletter *The Hardwood Herald Quarterly,* informs the hardwood industry about HMA activities. A page from this publication in shown in Figure VII-1.
- Education on the selection and care of hardwood furniture, cabinets, and architectural woodwork. The program aims to inform future specifiers of home furnishing products on the unique characteristics of solid hardwoods. Supplying literature for classrooms and guest lecturers from the hardwood industry, the HMA includes brochures, videos, and a hardwoods poster in a hardwood education kit for college home economics and interior design instructors.
- Displays have been extensively utilized in the HMA traveling home show display that tours home shows throughout the country. The shows generate thousands of inquiries on the use of hardwoods in the home.
- Demonstrations of hardwoods serve to whet the public appetite for hardwoods.

FIGURE VII-1
A Page from the Hardwood Manufacturers' Association Newsletter
Hardwood Herald

HARDWOOD HERALD

F A L L 1 9 9 0

TODAY · TOMORROW · SOLID HARDWOOD · FROM · THE · USA · FOREVER

INSIDE THIS ISSUE

HARDWOOD *Expressions* Grows For Fall

The Fall 1990 issue of Hardwood Expressions, the Association's national consumer magazine, includes four additional pages and will be distributed to more than 15,000 consumers in early October.

The Fall issue will be 20 pages compared to the 16-page premiere issue which was distributed in July 1990. The size of the magazine has increased because advertising is now being accepted.

Advertisers benefit from reaching a highly targeted market through Hardwood Expressions because the magazine's demographics are targeted toward high-income households that have either recently purchased or plan to purchase hardwood products in the near future.

Another way that manufacturers and dealers have utilized Hardwood Expressions in their marketing efforts is by participating

continued on page 2

HMA Launches Fundraising Campaign

HMA is preparing to launch its most ambitious fundraising campaign to date by delivering a special message to the top executives of a targeted group of solid hardwood products companies.

That message, according to Stephen H. Conger, Promotion Committee Chairman, is that the industry must increase its investment in promotion when business conditions are soft, especially in light of inroads gained by wood substitute products.

continued on page 3

Home Show Continues

CONTEST HELPS FURNISH HOMES

As the HMA Home Show display rounds out its seven-city 1990 tour, more than 2.4 million consumers will have visited the home shows in which HMA exhibits.

This year the exhibit has completed stops in Boston, Detroit, Indianapolis, Mobile, Anaheim and the Lake States Logging Congress, sponsored by the Lake States Association, in Green Bay, Wis.

The remainder of the exhibit schedule for 1990 includes visits at the HMA Production Meeting in Madison, Wis., Oct. 3-5; and the Virginia Interior Design Show at the Virginia State Fairgrounds, Richmond, Va., Oct. 26-28.

continued on page 3

1

The "solid hardwood apartment" of noted architect Michael Rubin in New York City is featured as one demonstration of the beauty of hardwoods. Four inch maple plank flooring throughout sets off custom maple cabinets and furniture. The apartment's setting with a view of the East River and downtown lights of New York City provides a dramatic background.

Solid United States cherry, maple, walnut, poplar, and oak hardwoods are featured in another demonstration, the "Busy Woman's Dream House," designed and built under the sponsorship of *Family Circle.* Maple planks create a striking trussed ceiling above the dream house living room area. Hardwood floors, cherry inlaid with walnut and maple borders downstairs, cherry stairs secured with walnut pegs, poplar architectural columns, and similar solid wood are featured.

❑ A distinctive logo, "solid hardwood from the USA" with the words, "Today, Tomorrow, and Forever" carries out the theme.

❑ Point of Purchase (POP) materials include solid hardwood hang tags and cards, a hardwood trademark rubber stamp, an oak literature rack, and many pieces of literature for furniture and cabinet makers to distribute.

SELECTING A PROMOTION STRATEGY

The promotion strategy selected depends on the company's objectives.

A pulling strategy is often used to promote a new product or ex-pand sales of an existing product. The pulling strategy focuses on stimulating customer demand for a product: A customer reads an advertisement or receives a form of sales promotion about the product and asks the retailer for it. The retailer, in turn, orders the product from the distributor. The product is pulled from the manufacturer through the distribution channels to the customer. Advertising and sales promotion are the usual tactics for a pulling strategy.

A pushing strategy stimulates sales by promoting the product to distributors and other members of the marketing channel. The pushing strategy relies on trade discounts, cooperative advertising allowances, and similar incentives to motivate wholesalers and retailers to spend time and effort in promoting these products to consumers.

Push and pull strategies are often used simultaneously.

The promotion strategy that helps increase sales integrates personal selling, advertising, and sales promotion into one blended promotional mix. Achieving that integration depends on how thoroughly a company understands its potential customers. Ten questions for analyzing potential customers as a basis for designing a promotion strategy are listed in Figure VII-2.

FIGURE VII-2
Ten Questions to be Answered in Designing a Promotion Strategy

1. What is the target market to be reached by the promotional strategy?
2. Who are the persons and their positions in each customer target company to whom promotion should be directed?
3. What is the size of the market and number of customer firms to be reached?
4. What are the factors that influence the buyers in the targeted companies?
5. What are the objectives of personal selling, advertising, and sales promotion in influencing these buyers?
6. What is the specific message to be communicated to the buyers?
7. What is the best media combination to carry the intended message?
8. What are the budget requirements for the promotional mix?
9. How effective is the strategy?
10. How effectively can the promotion reach potential customers?

Management's experience with the additional sales produced by promotion is evidenced by increasing expenditures on advertising and sales promotion. Modern institutions, including non-profits, government, and religious organizations rely on promotion for survival. And today's activist society, wherein every person believes he has a right to participate in decisions affecting his life, relies on promotion activities to gain public support.

PROMOTING NICHE PRODUCTS

Niche products, by definition, aim to appeal to a narrow segment of the market. The advertising theme reflects the product differential that specifically matches that niche's wants and needs. Selecting an advertising theme is relatively simple. Choosing the media in which to communicate the differential advantages of a niche product is more complex.

In marketing niche products, innovate or focus on whatever sets the product apart from others, and utilize the media that is most likely to reach potential customers.

> The successful regional chain of Izzy's Pizza Restaurants found its niche in upscale dining and a broader menu than that offered in most pizzarias.

Izzy's founders recognized the emotional issues of pizza crusts and sauce and developed a unique pan crust. Commercials made by one of the owners feature customers eating the pizza crust first. Heavily promoted through local cable television, Izzy's relies on customer opinions to maintain high quality. Sensing the trend toward low calorie, low cholesterol, Izzy also developed a pizza that satisfies appetites and health concerns.

Izzy's selected local cable television as the medium most viewed by the customers they were trying to attract. The company chose advertising as the best method to reach the largest number of potential customers. More important, Izzy's followed four basics of niche marketing:

- ❑ Find a "hole" in the market and devise a way to fill it. (The "hole" in the market was the upscale variety of pizzas. Izzy's filled it with quality pizza.)
- ❑ Develop one or two ways the product can stand out from the crowd. (Izzy's developed the pan crust.)
- ❑ Select the medium that reaches the target market and advertise why the product is better or fills a need other products don't. (Izzy's used television.)
- ❑ Listen to and act on customers' opinions. (Customers were asked to rate the quality of food and service on a written card. Replies were analyzed by computer.)

THE ART AND SCIENCE OF PERSONAL SELLING

Personal selling has been a part of business for thousands of years. Whenever people transmit ideas to each other, they are engaged in personal selling: The mother encouraging her offspring to have good manners is engaged in personal selling, young men and women on dates spend much effort selling themselves, and any person running for political office is in personal selling. Some people are far more persuasive than others in face to face contact. The super salesperson has long been described as someone "who can sell refrigerators to Eskimos." Other would-be salespeople have been tagged as unable to sell anything, "even if the customer wants it."

Both industrial and consumer products can be promoted by personal selling when:

- ❑ Customers are geographically concentrated
- ❑ Customers require additional information and instructions to use the product or service
- ❑ Products are custom made or expensive
- ❑ Customers require special attention due to sales volume or complexity of product

The high cost of personal selling limits its use. Startup companies often cannot afford personal selling. At an average of four visits by a salesperson to complete a sale personal selling is usually more costly than advertising. Computer terminals, commission-only salespeople, and telemarketing are attempts to decrease selling costs and improve selling efficiency.

A survey indicated that almost 70 percent of industrial products are sold by personal selling in contrast to 48 percent of consumer products. Industrial products rely somewhat less on advertising than consumer products.

Consumer products use more special promotional activities. An attractive aspect of personal selling is flexibility and efficiency of personal selling in closing sales and providing feedback. But much consumer personal selling is in the hands of untrained sales personnel in retail stores who merely take orders, answer inquiries, make recommendations, and receive payments and pack products.

> The Body Shop, a retail store that features natural lotions for the body and natural cosmetics, has become a legend in the United Kingdom for its special type of merchandising. All sales clerks are trained to consult a huge company reference book called The Product Information Manual, which tells everything there is to know about the products—where they come from, what is in them, how they are made, and what they can be used for. The manual has helped salesclerks earn great credibility with customers. Promotion through personal selling has been the marketing strategy that is responsible for the outstanding success of the $100 million business of The Body Shop.

Volumes have been written on how to be a productive salesperson, but in the last analysis personal selling does not follow a series of rigid rules; many business executives believe it is an art. Whether good salespeople are made or born is a popular business luncheon topic.

Modern, professional salespersons are a far cry from the early peddlers and owners of stalls in the town square who hawked their wares to potential customers. Today's sales personnel help customers select the right product, rather than sell whatever is available. The interests of the buyer are closely tied to those of the seller.

The Stages of Personal Selling

Certain protocols and procedures in personal selling make a good salesperson better and help develop sales ability. Closing the sale is one of the most difficult functions of the salesperson. The seven stages in closing a sale are shown in Figure VII-3.

FIGURE VII-3
Seven Stages in Closing a Sale

1. Identifying and qualifying potential customers, determining that the prospect has the authority and money to make the purchase. This stage is often frustrating and slow to pay off, especially in major industrial sales, which may take years to consummate. Communication between the marketing and sales departments is often lacking and sales personnel are not fully informed about the target customers or segments selected by market managers.
2. Approaching the prospect. Here information about the potential customer's business and personality helps the salesperson decide whether to take a low-key or an aggressive approach.
3. Presentation of the sales message. This refers to the product features and how they relate to solving the customer's problems. If the product has been advertised, the salesperson repeats the differentiation benefits and any positioning features at this time.
4. Demonstration, where appropriate, makes an effective sales presentation. In presenting a new piece of ready to assemble (RTA) furniture to a purchasing agent, demonstrating the ease with which it is assembled may clinch the sale. Demonstrations should be carefully planned so they live up to their claims. Having to apologize for a missing screw in an RTA demonstration projects a negative image of the product and the salesperson.
5. Handle objections by presenting further information. The customer may voice his objections after or during the presentation.
6. Closing the sale. Closing the sale is for many salespeople the most difficult part of the job. Some salespeople use customers' fears, pride, or other emotions as the key to closing. Others have an assortment of special services or quantity discounts in waiting to achieve a closure. Salespeople who know their product and their prospects close sales by offering alternative decisions—buy now for cash, or later for financing; buy this model for quick payoff, that one for long term durability. Telling the prospect that the product may not be available later or forcing the prospect to make some decision are other techniques.
7. Follow up every sale and contact. Sales follow up provides an opportunity to rectify problems and assure customer satisfaction.

The Sales Meeting

The annual meeting of sales personnel called by management to whip up enthusiasm for selling the company's products is a time honored event. For one salesperson or one hundred, the sales curve rides on how well management conveys information about and expectations for the product. The old type of annual sales meeting featuring rah-rah, scotch, golf, and parties is on its way out, according to many sales managers. However, there is greater urgency than ever for the boss to get together with the firm's salesforce. For one or a few salespeople, it costs management little to find a pleasant little retreat and spend a day going over new products, listening to feedback from the field, and developing realistic sales goals. Arranging the large sales meeting at a well-known resort is a job for a professional meeting planner. Small or large, the sales meeting will better stimulate sales if it focuses on the sales force rather than management and meets these five criteria:

❑ Has a clear purpose

Today's salespersons are likely to be different from the those management grew up with. More have degrees. More are women. More have long range plans for their own future and are interested in how the company's plans help them. If the purpose is to introduce a new product, consider substituting video tapes, cable TV programs, teleconferences, local meetings and seminars to transfer the knowledge just as effectively and at a lower cost.

❑ Is well planned

A company's sales meeting is as important as anything else on its schedule and deserves to be planned to the last detail.

❑ Brings all levels of the company together

Opportunities for communication are lost if executives sit at one table and the sales staff at others. The objective of the sales meeting to stimulate a team effort is difficult to achieve if the participants don't get to know each other better.

❑ Weighs carefully whether speakers engaged from outside the company are motivational or informational

The old fashioned motivational speaker may not be as effective in accomplishing the meeting objectives as an expert bringing new insights that may affect sales. If the expert is witty and amusing, so much the better.

❑ Meeting goals and company goals are closely linked

The sales meeting can't be isolated from the company condition. It should relate to what's happening in the industry, the nation, and the world. If product quality has been a concern of the sales force, specific company plans for improving quality sends the sales force forth with new vigor. When the economy goes through one of its inevitable declines, meeting budgets are first to be slashed.

Companies that rely on personal selling can barely afford to curtail meetings in the face of greater difficulty in maintaining a sales position. A declining economy simply triggers tightening every facet of the meeting budget, not eliminating it.

ADVERTISING

Advertising is mass communication directed to a large number of potential customers. In advertising, a firm pays a medium—newspaper, magazines, radio, TV—to carry its message and identifies itself with the message. Dollar costs are high (one 30 second TV ad could run well over $150,000), but per viewer or listener cost is low. Advertising in the mass print media is likewise costly. Companies that market to target segments of the population, rather than mass marketing, waste fewer advertising dollars on viewers who are not prospective buyers of the product.

Advertising as an industry began in the United States in 1841. In the intervening years, its focus has broadened from selecting the best media for the product to understanding customer behavior and creating a favorable company image.

Polls have shown that the public's confidence in business has declined from a little over half to less than one-third. Creating a favorable company image as an objective of advertising has acquired an urgency in the past 30 years. All forms of promotion, including personal selling, go into the mix for building a favorable company image: one bad personal experience with customer service personnel can undo thousands of dollars of advertising and publicity efforts.

The Media

The diversity of media, ranging from radio and TV through newspapers and magazines, to direct mail and outdoor media offers advertisers a wide choice.

The media is undergoing a technological evolution and by the year 2000 new media will have sprung up to tantalize advertisers. Computer voice machines and computer video hooked up in interesting new ways, will expand methods to communicate beyond traditional media. Linkage via telephone with personal networks will speed and direct messages to the target customer. Fax machines have already created a new advertising medium and cable TV has opened up broadcast advertising to local and smaller companies. Innovative companies have an unprecedented opportunity to experiment with these new forms of communication.

The most important advantage of advertising is that the company placing the advertisement has control of the content, timing, size, and length of the message. However, a mistake in selection of media can eat up advertising dollars with the efficiency of a paper shredder.

Any medium selected should be subjected to these same tests:

- ❑ Will it reach the targeted public?
- ❑ Is the reach sufficiently large in proportion to the targeted public?
- ❑ Does the medium have credibility with the targeted public?
- ❑ Is timeliness essential? What are the lead time requirements of the medium?
- ❑ Does use of the medium have a possible negative consequence, i.e. a forest industry alliance advertisement or publicity on forest management plans in an environmental magazine?

Each medium has its own advantages and disadvantages. Traditional media includes:

(1) Daily newspapers

Newspapers offer flexibility, coverage of the desired area, tie-ins with sales promotions, and the ability of the reader to refer back or clip the advertisement. Advertisements in daily papers help build company and product prestige. The disadvantages are that newspapers are quickly discarded and readers have limited time to spend reading. (Reading time has been found to vary with age groups, education, and occupation.) For expensive wood products in which quality and detail are differential benefits, reproduction in daily newspapers is too poor for consistency with the quality image.

The newspaper advertisement from a local newspaper shown in Figure VII-4 promotes fencing. The advertisement features prices of options and available services.

(2) Weekly newspapers

In communities served by weekly newspapers, coverage tends to be more intimate and more personal, but the percentage of readership is higher. The weekly papers have a smaller number of pages and advertisements are usually less expensive.

(3) Television—national networks, national and local cable

The fastest growing advertising medium is television. But relatively few forest products manufacturers use national television "spots" or even local advertising spots. However, building supply stores, home

FIGURE VII-4
A Newspaper Advertisement to Promote Sales of Fencing

center stores, and other retailers often use locally developed commercials. High impact in generating product and company prestige, repetition, and flexibility and the ability to reach millions of viewers make TV a dramatic medium, although the millions of viewers may contain very few prospective customers who are reached at exceptionally high cost, with a message they may not remember for more than a few seconds.

Network television offers new innovations to compete with cable and independent stations that provide advertisers a choice they didn't have some years ago. One innovation is tie-ins for specific brand ads. Columbia Broadcasting System (CBS) broke its long standing rule against mentioning brand names on TV programs by using a Coke machine as one of the props in a program. Other tie-ins involve using TV to allow viewers to enter sweepstakes or contests. The innovation uses toll free telephone numbers as a method of issuing coupons and rebates.

The half hour and full hour commercials in talk show or documentary formats are an exploding business. Diets, cosmetics, fitness, and get-rich-quick schemes are popular topics for these extended commercials. The programs are made possible by the abolishment of commercial time restraints in the 1984 deregulation of the broadcast industry.

The forest products industry has not made notable use of these long commercials, but it does benefit from public service programs, such as "This Old House," which demonstrates do-it-yourself remodeling projects. An opportunity awaits for the furniture and cabinet industry to cooperate in long commercials about how to buy furniture and the construction details of furniture and cabinets. Home center stores have the opportunity for half hour commercials on do-it-yourself projects.

(4) Direct mail—postcards, leaflets, sales letters, printed product technical literature, booklets, catalogs, house organs, and fax messages

Costs of direct mail range from the relatively minor cost of computer printed sales letters to the major cost of catalogs. Computer software permits personalizing each mailing piece, and the mailing list can be screened to reach the most likely potential customers in the target market. The flexibility of direct mail formats increases daily as new desktop publishing techniques are made available for small offices and local desktop publishers establish businesses.

Many companies sell targeted mailing lists based on zip code, age groups, type of business, hobbies, and other groupings.

(5) Magazines

Magazine publishing today is distinguished by an increasing number of magazines focused on special interests, such as women, yuppies, small businesses, racial groups, specific industries and political viewpoints.

As the market has fragmented into hundreds of pieces, the media that people read has likewise been sliced to reach smaller and smaller audiences.

Magazines aimed at mass audiences have experienced declining readerships. Targeted magazines such as *Black Enterprise* and *Wildlife* have replaced national mass publications such as *Life* and *Look.* Specialization is the trend. This is good news for advertisers. They can fashion advertisements and choose publications that reach their market. The readers of these specialized magazines constitute an active market for products designed for them.

Magazines lend themselves to quality reproduction, adding a seductive quality to items illustrated. Magazines are kept longer than newspapers; some of the more expensive magazines are collector's items displayed on the coffee table in the living room.

(6) Trade journals

Trade journals are in a class by themselves as vehicles to reach potential industrial customers. Suppliers of wood furniture components reach customers directly by advertisements in *Furniture Design & Manufacturing* and other trade journals aimed at furniture manufacturers. Producers of custom cabinet doors reach cabinet makers with advertisements in *CabinetMaker.* Manufacturers of furniture frames for upholstered furniture reach customers in *UDM/ Upholstery Design & Manufacturing.* Producers of lumber packaged for do-it-yourself customers advertise in *Home Improvement Center.*

For advertising industrial products, trade journals have no peer. Prospective buyers moving toward a buying decision read the trade journals in their industry. Advertising in these journals creates awareness and preference for the product. When the salesperson makes contact to close the sale, the advertisement has played a role by:

❑ Providing information on the company and its products prior to the salesperson's call

- ❑ Creating a favorable company image to inspire customer confidence
- ❑ Reaching technical or executive personnel who might not be contacted by salespersons but who influence buying decisions
- ❑ Reducing the costs of selling
- ❑ Improving the buyer's familiarity with products
- ❑ Positioning the product to meet the customer's need

(7) Radio

Radio is infrequently used for advertising wood products. The temporary nature of the message, and the fragmentation of the medium into 20 or more stations a listener might hear in any one area are not balanced by lower cost, immediacy, flexibility, or other advantages. Radio is a medium that might be used effectively by wood products firms to establish a perception or correct a misperception in a positive manner. An example is a series of radio messages on the protective value of the permanent wood foundation in an area that has just been scared by the rumors of radon entering homes through the soil.

Radio advertisements are much easier to create than TV advertisements and can be produced rapidly and at less cost. Radio is useful for lumber yards and home center stores for advertising sales, specials, and manufacturers workshops.

(8) Business papers

Like trade journals, papers published by business organizations reach people involved in a specific industry. Examples are the papers published for the furniture trade, the contract construction business, and the lumber trading business.

(9) Outdoor advertising

Billboards, painted bulletins, signs on the walls of buildings, electrified or animated signs, and hot air balloons with messages are all part of outdoor advertising. Infrequently used for wood products, outdoor signs are effective on the walls of buildings of manufacturing plants or building supply dealers. Outdoor signs communicate brief messages repetitively but their location and size are limited by federal regulation (the Highway Beautification Act of 1965) and often by local ordinances.

(10) Directories—telephone, state, trade

Advertisements in telephone directories are relatively inexpensive methods of communicating quick messages about companies and their products. Particularly useful for local services and consumer products, telephone directories have the advantage of frequent consultation and a degree of permanence that exceeds other media. Most states publish directories of manufacturers in which companies may be listed at no charge. These directories are often consulted by prospective buyers and have the advantage of identifying categories into which products and services fall. Trade directories such as Thomas Register of Manufacturers, Sweet's Catalog, and the Directory of the Forest Products Industry—published annually by Miller Freeman—provide a handy reference to manufacturers of categories of wood commodities and wood products. Listings in such directories, usually without charge, reach the segment of the market ready to purchase the product.

(11) Throwaway flyers

With the advent of desktop publishing, throwaway flyers should become more popular as a medium for advertising. Inexpensive, flexible, readily retained by prospective buyers, these advertisements carry information to interest the customer. Distribution of these flyers at county and state fairs, trade shows, and exhibitions attended by persons who might be interested in related products (such as home shows), anywhere there might be a flow of traffic, costs little for the impact possible with flyers.

Types of Advertisements

Advertisements play specific roles. With products, they may inform, persuade, or merely remind prospective buyers that the product is available for purchase. Their role at any given time is related to the product lifecycle. In the introductory phase, advertising informs prospective purchasers about a new product or a new service. During the growth phase of the cycle, advertisements aim to persuade potential customers to become buyers. In the mature and declining stages, the goal of advertising is to remind the public that the product is available and competitive.

When a company promotes a new product, such as an engineered I-beam, it uses informative product advertising. When that same company wants to promote goodwill for the forest industry, it places an institutional advertisement.

Institutional advertising promotes goodwill for a company, an industry, an organization. Or it may promote an idea or a philosophy or a cause. Institutional advertising helps build a desired image. Cities or localities attempting to build an image for industrial development often use institutional advertising. An organization wishing to remind readers of the value of its activities, uses reminder institutional advertising. Nuclear energy, the forest industry, the coal industry typically use reminder institutional advertising to win goodwill.

In the Spring of 1990, retailers and manufacturers in the Home Furnishings Council (HFC) joined together to sponsor a series of institutional advertisements emphasizing the comfortable home as a haven. Robert W. Nightengale, Jr., HFC co-chairman remarked: "For the first time, the home furnishings industry is speaking directly to the consumer with a unified voice." To support the campaign, HFC prepared a home decorating primer. The primer is mentioned in the ads with a toll free number to find out where copies are available.

A firm may accomplish its advertising objectives by using *comparative advertising,* making direct comparisons with well-known competitive brands. "Nationwide, more Coca-Cola drinkers prefer the taste of Pepsi," advertises Pepsi-Cola. The comparative strategy may be used to advertise solid wood products as compared with composite products, or vice versa, as in Figure VII-5. Advertisments may use *celebrity testimonials*—big-name personalities who help win product recognition.

Advertising industrial products, as with consumer products, aims to bring about some change in awareness, or knowledge, or attitude in a target market.

Whatever its objective and style, following six guidelines will help create advertisements that work:

❑ Select a theme. Advertising communicates the differential advantages of the product to the potential customer. Appeal to the customer with a statement that tells how the product meets a need. Describe how it is used; how it can save money, time, or work; or how it can improve a person's chances of success. Some themes are based on fear, such as risk or loss if the product is not used. The theme reflects the positioning strategy selected to promote the product.

❑ Focus on persuasive impact. Persuasive impact is the ability of the advertisement to evoke a reaction. Consumer product advertising usually attempts to produce an emotional response, rather than merely convey information. Evoking emotion, moving the customer to see himself or herself in a commercial, results in 50 percent higher retention than merely providing information. Consumer advertising is often used to bring a potential customer from awareness to actual purchase.

The furniture industry is changing from bland ads in decorating magazines to persuasive advertising calculated to tug at the heartstrings. La-Z-Boy's advertisements previously showed chair frames stripped of their upholstery

FIGURE VII-5
An Advertisement Comparing a Composite with Solid Wood

and featuring hardwood and solid foam. The new advertisements sell warm feelings with the theme "nobody has rocked more babies than we have. In a TV commercial a young father paces the floor with his new baby and settles down in his La-Z-Boy rocking chair to cuddle the baby.

The HFC is turning out creative ads to reposition furniture as a fashionable lifestyle purchase, rather than a lifetime investment.

❑ Carry out the marketing strategy. The advertisement links the producer with the customer. Advertising communicates the manufacturer's positioning statement to the potential customer. If the advertisement is not consistent with the strategy, the strategic planning and the advertisement are wasted. An effective advertisement immediately identifies the company.

❑ Reach a sufficient audience to be cost effective. "Reach" is the advertising term used for the number of viewers or readers who are exposed to the message.

❑ Deliver the message with sufficient frequency to accomplish its objective. "Frequency" is the advertising term that indicates how often the message is delivered. Use newspapers for the message that needs to be delivered daily. Advertising in telephone directories delivers the same message all year. The manufacturer of kits for constructing wood decks may enter into cooperative advertising with a local building supply dealer to place frequent newspaper ads that run during the construction season.

❑ Select a medium that assures the message remains with the audience as long as necessary to accommodate the buying mode. "Message permanence" is another advertising term. This term indicates how long an advertisement remains with the audience: a TV ad may last only 5 seconds and be forgotten by the time the consumer is ready to buy. An outdoor ad is visible for months. To evoke a high emotional response, choose TV. For reminder advertising, use the outdoor ad. For calling attention to a special event, use newspaper advertising.

PUBLICITY

Publicity, like advertising, is a form of mass communication. It differs from advertising in several significant ways, primarily in that the media is not paid for presenting the message, and the firm is usually not identified as the sponsor of the message. Product publicity is part of product promotion strategy. Publicity can supplement a low budget advertising campaign by carrying product news items that are not paid for in the print or broadcast medium. As a technique to create an awareness or interest in the product, publicity is often more effective than advertising in promoting products and services.

Publicity is one of the tools of public relations. Public relations encompass government affairs, participation in public endeavors, and similar activities. Public relations to protect and enhance the prestige and image of a company or industry by gaining favorable public opinion has assumed greater importance in today's era of participatory democracy. The right of all people to participate in decisions affecting them—the basis of participatory democracy—has given consumers a place in board room decisions.

Public relations is a planned effort to influence public opinion through communication, and through relationships with the various publics who influence legislation. It is also an effort to influence the business climate in which companies operate. Publics can include employees, customers, stockholders, suppliers, the government, and other stakeholders in the company. Public relations are sometimes handled by the public affairs departments of larger companies.

Planned publicity includes news releases, open houses, ceremonies, plant openings, speaker's bureaus, and press tours. Publicity makes news happen. A skillfully planned event is a newsworthy happening.

Principles of Public Relations

The book, "Public Relations and Communications for Natural Resource Managers" by James R. Fazio and Douglas L. Gilbert, (1981, available from Kendall/Hunt Publishing Company, Dubuque, Iowa) lists seven principles of public relations:

- Every action makes an impression
- Good public relations is a prerequisite of success
- The public is actually many publics
- Truth and honesty are essential
- Aggressive action is more effective than defense
- Communication is the key to good public relations
- Planning is essential

Using Publicity Releases to Promote Products

Publicity releases to trade journals are a low cost method of generating interest in products. Trade journals welcome new industrial product announcements as part of their reader service. The publicity announcement provides an opportunity to describe product differential advantages and to position products to gain buyer preference. If a company has instituted a new quality control, zero defects program, the publicity story can elaborate on the advantages for the industrial buyer. Publicity releases, like advertising, accomplish their objectives by carrying out the selected themes of product differentiation and product positioning.

Space in the printed media is limited and a publicity release for products will win a share of space only if it is newsworthy. A newsworthy publicity release about a product, or an idea newsworthy story, should be carefully scrutinized for:

- ❑ Timeliness. Is it tied to an event, a holiday, or a season?
- ❑ Reader interest. Draw on suspense, sympathy, uniqueness, progress, or the age and emotions or basic interests of readers. A publicity release on a unique wood clock could describe its unusual qualities. The story of a well-known person using the product lends interest. A wood product manufacturing plant may present a new product to the mayor of the city, with the press present.
- ❑ Relevance. Is it related to the area or community served by the local newspaper or popular magazine?
- ❑ Importance. Does it tell readers something important? For example, the company has added extra employees to handle the demand for a wood product as Christmas gifts.

Publicity Opportunities

Publicity is truly a golden opportunity for the company operating on a shoestring. A number of publicity opportunities are low in cost and high in effectiveness.

One low cost publicity activity is organizing a company speaker's bureau and offering to talk about a topic of general interest related to the company or its products. Service clubs, chambers of commerce, church groups, and other organizations have a constant demand for speakers. The persons selected to speak for the company need not be polished speakers. A friendly, informative talk can transmit the public relations message.

Writing technical articles for trade journals is a method for reaching purchasers who make technical decisions.

Providing articles featuring a concept or product as part of hints on home decoration or gardening for the home and garden sections of daily newspapers can result in very positive impact.

An example of a powerful publicity campaign is the "bedding revolution" of the late 1980s. Americans who once bought sheets at white sales and thought bedspreads were a bed covering are now spending hundreds, even thousands, of dollars to decorate their beds with designer linen and special mattresses and mattress pads. Carol Colman, vice president of Inferential Focus, a New York consulting company that spots trends, explains the evolution toward decorating a room that guests hardly ever see as an opportunity that people seize to control their own lives in a world where so much seems out of control.

Providing a public service is public relations that builds good will and enhances the prestige of a company.

One way to build good will is to give an award and publicize it through the media. The local wood producers association can present a plaque to the company that contributed most to local employment or to the economy. A description of new products can be included in the award publicity. A firm may introduce a new product at a special event, an art exhibit or an exhibit of tropical wood species for an upscale wood product.

FIGURE VII-6

Turning Coffee and Doughnuts into a Publicity Coup

It is not unusual for a company or distributor to serve coffee and doughnuts to customers. That's good customer relations, especially if customers are in a narrow geographical area and occasionally visit the manufacturing plant, or lumber yard, or store.

The same friendly activity can be turned into a publicity coup that calls the attention of the public to the company and its product by establishing the custom of serving free coffee and doughnuts (or low cholesterol cookies) once a week, preferably on a usually slower day.

Next, send letters to customers and potential customers informing them that on the chosen day free coffee and doughnuts will be served, and they will learn something to help them increase their sales. Include that news in any advertisements carried in local or regional media.

Group the information opportunities into a series of topics, to be presented over a number of "coffee days." Improved methods for constructing a patio deck, or what's new in safe wood preservatives are two example topics. Select suppliers or experts to present the information and keep the sessions short. Handouts can supplement information.

Now the company has the ingredients for a news story. Write a publicity release with an attention catching headline such as "Experienced Remodelers Get Up Early to Learn What's New in Decks (or Woods or Preservatives). Learn to Serve Public Better."

Contact the local newspaper and invite them to take photos and cover a session. They may wait for a slow news day, but sooner or later the article will appear. If the area has local TV coverage, they may be interested in a TV program.

Attendees will be pleased to be part of a news event. The continuing program will draw the attention of potential customers who may want the distinction of joining the group. The company and its product or service gain visibility. The cost to the company is the price of a few doughnuts and coffee—a bargain in public relations!

Another way might be to sponsor safety advertisements as an individual company or in a cooperative ad with other companies. The advertisements may be for fire prevention, against drunk driving, or for other causes.

Many companies sponsor local baseball, softball, other sports activities or cultural events such as opera, symphonies, and art exhibits.

A lumber company or lumber yard can earn local good will by donating lumber to a civic building project.

Valuable publicity can be based on an activity as simple as serving coffee and doughnuts to customers. How to turn this simple act of hospitality into a publicity coup is described in Figure VII-6.

SALES PROMOTION

Sales promotion consists of activities—other than advertising, selling, or publicity—designed to stimulate customer interest and purchases. Sales promotion assists dealers and other middlemen in making sales, helps to attract customer traffic in retail stores, and builds and maintains brand loyalty.

The most widely used method of promoting industrial sales is the product catalog, which prospective purchasers consult often when they are prepared to buy. From a two sided brochure to a large, bound volume, well prepared catalogs are a powerful sales promotion tool. A precise knowledge of what influences buyers of a product helps in compiling a catalog. The book "How to Build and Distribute Your Next Catalog," is available from Thomas Publishing Company, One Penn Plaza, New York, New York 10001, telephone (212) 695-0500.

Trade Shows are the second most widely used form of industrial sales promotion. A trade show organized by a trade association or professional group displays products to attract distributors. Home and recreation shows offer consumers a one stop inspection opportunity to learn about new products for the home, home care, recreation products, and health aids.

A trade mission is a type of trade show that is gaining popularity in promoting exports. The trade mission may be sponsored by the United States Department of Commerce or Department of Agriculture or by the international division of the economic development department of a state. Manufacturers are given an opportunity to exhibit their products and distribute literature. The international staff at either the federal or state level uses its expertise to help set up appointments, make travel arrangements, and generally pave the way to making sales contacts.

State trade missions also participate in major industrial trade shows in the United States to promote the wood products made in their states.

They are also used to inform wood products manufacturers about their local timber resources.

At a trade show a company parades its wood products before potential buyers and:

- ❑ Meets potential customers
- ❑ Accumulates a mailing list
- ❑ Introduces new products
- ❑ Demonstrates products
- ❑ Locates new representatives and dealers
- ❑ Attracts new personnel
- ❑ Receives feedback from customers and potential customers on product utilization
- ❑ Enhances the company image
- ❑ Gathers information about competitors
- ❑ Sells products
- ❑ Generates sales leads

Trade journals for each industry list the trade shows for that industry. For example, Pallet Enterprise includes trade shows of interest to pallet manufacturers.

According to Richard Murphy, director of marketing for Giltspur, Inc, an exhibit/marketing and design firm in Itasca, Illinois, most exhibitors muddle through trade shows. In 1989, companies spent more than $21 billion to exhibit at more than 9,000 trade shows, but 56 percent didn't train their staffs, and about 78 percent didn't promote their exhibit prior to a show. The typical exhibitor, it is said, misses 40 percent of the potential prospects at a given show.

That 40 percent can be reached and more of the 60 percent usually reached can be converted to customers by attention to the suggestions in Figure VII-7.

Other sales promotion tools include point of purchase (POP) displays. Especially useful in marketing to the do-it-yourself customer, POP displays attract attention and inform the customer about the wood species, grade—if appropriate—use, price, units, assembly requirements and other details that may influence selection of the product. POP tags that carry price only are lost sales promotion opportunities.

New England Millwork Company's machine-carved, decorative hardwood mouldings command a higher price than ordinary utilitarian millwork. To

FIGURE VII-7
Reaching Potential Customers at a Trade Show

To reach potential customers at a trade show, follow these suggestions:

- ❑ Design an exhibit that awakens interest with new information, new benefits, or visual demonstrations
- ❑ Train salespeople for the exhibit and have enough, but not too many, salespeople to handle the booth. Several studies have shown that what attendees remember most about exhibits are the actions of the exhibit staff.
- ❑ Build interest and excitement about the exhibit by using free mailing lists to promote attendance. Send a series of mailings that focus on the company's show theme—introducing a new product, a new service—promising a sample or a gift for listening to the presentation. Salespeople can add to the interest by making personal calls inviting customers and prospects to the show. Advance advertising of the features and applications of the product exhibited in appropriate trade journals generates interest. Keep up the excitement by advertising in special papers published for the show.
- ❑ Create a show kit that can be left with customers and prospects before the show. The kit can contain admission tickets, promotional material, and an invitation to visit the exhibit.
- ❑ Invite press coverage if a really new and exciting product is being introduced, or the company has an innovation to an existing product. At the least, submit press releases about the new product before the show.

induce retailers to handle their moulding as a higher priced specialty product, New England Millwork utilized POP displays and tags. The brand name "Beauty Mark" was imprinted on the back of each piece of moulding. In addition, the tag identified the species as a status material, another illustration of singularization. The species was described as "made from genuine walnut from Southern Indiana," or "made from premium Massachusetts red oak." A display unit header showed how to apply the product and provided hooks to display the tagged samples.

Supported by skillfully crafted publicity, the POP campaign succeeded in gaining recognition for "Beauty Mark" as a product of greater value than ordinary millwork.

Brand Identity Through Sales Promotion

Brand identity has become a critical competitive edge for all types of products. Sales promotion creates brand identity with a combination of visual elements, including the name or logo, color combinations, symbols, package configuration, and shapes.

> A large manufacturer of computer printers changed its selling focus from computer equipment makers and distributors to industrial users. Name recognition was unimportant under the old distribution method, but it was a significant factor in selling direct to industrial users. The company accomplished its objective with an ad campaign aimed at trade and general business publications to build recognition of its name.

Conventional wisdom used to hold that "house brands"—those identified with a particular store—appealed to either downscale shoppers looking for bargain prices, or upscale shoppers who compared labels. In the 1990s, the general population has swung to house brands in its quest for dispensing with unnecessary expense.

Specialty advertising uses small items printed with the advertiser's name, address, and advertising message to reinforce sales strategies. Most people have pens, pencils, and other items imprinted with a company logo.

> One wood products firm turns out thousands of small souvenir bread boards to advertise its cabinet line. The bread boards are used as favors in conferences, banquets, fairs, and other events where there is little competition from similar products. The bread board is especially effective as something of value to the customer that may be retained as a reminder of the product. The effectiveness of specialty advertising has been questioned, but according to an A.C. Nielsen survey, buyers are more likely to purchase from firms using specialty advertising.

Technical bulletins and sales kits for dealers promote products by making information available to customers.

Samples, products distributed free to gain interest in new products and obtain future sales, may be distributed door-to-door, by mail, via demonstrations, or in packages containing other products. Although samples are more appropriate for convenience or shopping goods, there is much room for innovative use by wood products companies. A wood product that features ease of assembly may be sampled with a leaflet including assembly instructions with lumber purchase statements or in packages of hardware.

Coupons offering a price reduction on the next purchase of an item at a retail outlet, while infrequently used for wood products, offer the opportunity to encourage sales. A retail lumber yard providing fencing kits, for example, may include a discount coupon for the next 8 feet of fencing. Decorative panels sold in a home center store might include a discount coupon for matching moulding.

Rebates, in which the customer obtains cash back for a purchase or trading stamps given with each purchase exchangeable for merchandise, are another method of sales promotion.

Premiums, items given free with products or services, are considered a prime sales motivation. An annual premium/incentive show might provide a showplace for new and improved premiums.

Coupons and rebates are also useful ways to test the results of advertisements.

> Some marketing managers worry that offering more and more coupons, bigger and bigger rebates, and sweepstakes prizes reaching into millions of dollars are resulting in consumer addiction to sales promotions. It is easier to give customers a coupon for a few cents off than to build brand name loyalty. The research firm of Donnelley Marketing reported that in 1986 companies spent close to $100 billion on these promotions, amounting to 66 percent of all marketing dollars. Although sales promotion is usually cheaper than advertising and easier to measure, heavy use of coupons appears to be turning many retail brands into commodities that compete on price. Some large consumer product companies are trying to get away from reliance on coupons and into value-added promotions that build a brand image.

HIGH IMPACT MARKETING

Marketing programs often take several years to show results, taxing the patience of management. High impact marketing produces quick results and has high visibility early in the marketing program. The results of the high impact program may be modest, but so are the costs. High impact marketing is customer focused and helps employees feel good about themselves and their company. High impact marketing has three advantages, as shown in Figure VII-8.

> High impact marketing doesn't dent the promotion budget, and is effective in stimulating sales and employee morale. Its major component is change, and its impact can be attributed to what The Economist terms "The iron law of revolution: The more you change, the more you have to change."
> Changing to an active customer focus revitalizes the organization and provides a background for promotion efforts.

ALTERNATIVE PROMOTION TECHNIQUES

A tight budget doesn't have to limit promotion. Another way to communicate with customers about products is through demonstrations. Demonstration clinics on how the product can be used or on how easily instructions can be followed are particularly useful for do-it-yourself products, or for introducing new products to industrial customers.

> To promote sales of a new do-it-yourself kit for a handsome wood yard storage building, the Handy Company initiated a program of demonstrations. The unique kit included precut and numbered components, siding,

FIGURE VII-8
High Impact Marketing

High impact marketing has the advantage of requiring:

- ❑ Little time to plan and implement
- ❑ Low out-of-pocket costs
- ❑ Activities that improve service and customer relations

Use these eight simple steps to achieve high impact marketing:

1. Schedule a meeting with each customer and ask them how your can product help them, or how your company is doing. Most customers are pleased to be asked, and want to take advantage of the opportunity to speak their minds. Note and act on this customer feedback.
2. Interview each employee to find out what they think you can do to improve the company's product and service. Employees are pleased to be asked and frequently pass on suggestions they would otherwise keep to themselves.
3. Analyze the list of existing customers and note what they buy and when they buy it.
4. Prepare a company fact sheet, stating the products the firm makes and what they are used for. Include the three most important differential benefits of each product.
5. Learn, record, and analyze how new customers heard about the product or the company.
6. Alert receptionists and all front office personnel to the names and projected arrival times of customers and visitors. Prepare a list of the names of customers who may telephone, so that operators know who they are.
7. Keep an up- to-date list of customers' names.
8. Attempt to develop some type of clothing accessory that immediately identifies employees, such as a necktie, t-shirt, lapel pin, jacket, or where appropriate, a visor cap. A retail or wholesale establishment can encourage or require employees who deal with the public to wear company uniforms, or at least a uniform blazer with a company logo. Uniforms have been found to improve employee morale and to develop a positive company identity.

shingles, and hardware for the storage building. Handy Company positioned its yard storage kit with the advertising promise that anyone could easily erect it in a weekend. But sales lagged. Many consumers did not believe they could erect an 8 foot high, 8 foot wide, and 8 foot long building in a weekend.

To persuade potential customers that construction really was that simple, Handy induced retailers to have storage building raising demonstrations and clinics in their parking lots. Customers could actually participate in the assembly and learn how easily the detailed instructions could be followed. The demonstrations were convincing and sales exceeded expectations.

An alternative to personal demonstrations is the video taped TV demonstration, which can be set up next to a POP display.

Demonstrations have been workhorses in opening up global markets for American and Canadian products.

The construction of an apartment building erected by Holland's largest construction firm using United States timber frame products and technology resulted in permits to build four-story, wood-frame apartments in the city of Amsterdam. The permits marked a breakthrough in acceptance of new building codes by the Netherlands. The demonstration was part of the 1989 Urban Housing Summit Conference cohosted by the American Plywood Association and the Foreign Agricultural Service of the U.S. Department of Agriculture. The conference attracted urban planners, builders, government officials and designers from 11 nations.

A Canadian Western Red Cedar exhibit in the United Kingdom National Garden Festival demonstrated outdoor furniture and garden structures made of Western Red Cedar. The Canadian Council of Forest Industries (COFI) sponsored the show, which included cedar trees and their utilization. Responding to the intense movement pressuring British wood products consumers and manufacturers not to use woods from tropical rainforests and other potentially endangered wood species, COFI emphasized that Canadian Western Red Cedar is not an endangered species and that cedar trees are being replanted at the rate of ten million seedlings a year in British Columbia.

Direct mail can partially substitute for personal selling of industrial products when the product differential is too technical or detailed for an advertisement. Customers respond to follow-up direct mail when every letter describes another aspect of the same theme. Direct mail can result in a sale if it catches the prospect's attention and clearly defines the benefits, the products, or services he will enjoy once the purchase is made. A direct mail campaign should always tell customers where or how they can buy the product.

Writing a successful direct mail letter becomes easier with experience. According to Chuck Custer, a communications expert in Bellevue, Washington, the letter should begin with catchy, enticing opening sentences to grab the prospect's attention. To elicit a response, get to the point instantly and with strong emotional appeal. Say in clear, simple words how the product or service will fill the reader's basic human needs.

Local lumber yards offering immediate service or immediate help in solving problems may use phone book covers to keep their name, phone number, and service message readily accessible. Yellow page advertising serves the same purpose as phone book covers, and it has the advantage of often being the first line of inquiry.

A fact book is useful to promote industrial products in which formulae, tables, charts, and reference data play an important role in the purchasing decision. Fact books are sometimes offered free in special interest publications. Field representatives can turn a cold sales call into a meaningful contact by delivering the fact book in person.

Clear installation instructions can help sell a product in which the ease of installation may be a major selling factor. Installation instructions are a key promotional tool in do-it-yourself product marketing.

The power of packaging can't be underestimated. In addition to emphasizing the product benefits, packaging of a product can include advertising, installation instructions, information about the firm, brochures about related products, and even institutional advertising.

Listings in trade directories are usually free and provide product information to buyers. Most leading trade journals in the forest industry publish an annual buyer's guide. Listings in trade journals that cover the industries to which the company sells is a first step in promotion. It is an opportunity to tell potential customers that a company provides the products in the listing category. A moulding manufacturer, for example, not listed as a source of supply will probably be missed by buyers.

Using an 800 telephone number has become an inexpensive method of promoting products. Industry analysts estimate that toll free telephone services are growing at a faster rate than new telecommunications services. Toll free numbers are producing more sales than door to door selling ever did. AT&T surveys show that 86 percent of consumers would rather call an 800 number than write a letter to a company and 62 percent are more likely to do business with companies providing an 800 number. An analysis of a recent issue of a trade journal for the furniture industry showed that only one-third of the advertisements included an 800 number. Communications consultants say that companies not using an 800 number are at a competitive disadvantage. (Fax numbers, however, were included in more than 90 percent of the ads.)

Toll free 800 numbers are used to:

- ❑ Reach smaller customers who cannot be economically serviced by personal sales calls.

  Signal Thread Co., a manufacturer and distributor of commercial-grade sewing threads, more than doubled sales to smaller customers when it installed a toll free number. Signal President Steve Cooper says, " We're making sales we never made, reaching a market we never reached—all through the telephone."

- ❑ Handle mail order catalog calls. Lands' End, a catalog company with sales of $545 million, services over 5 million customers with an 800 number handling about 30,000 calls a day (80,000 during the Christmas season).
- ❑ Establish consumer hotlines
- ❑ Expedite service calls
- ❑ Provide customer support. WordPerfect Corp., marketer of the WordPerfect word processing program, provides intensive 800 number telephone support to help customers use their product. They spend twice as much on toll free customer support than on advertising, with more than 565 support people taking 13,000 calls a day.

Companies using an 800 number promote it heavily in advertisements, business cards, catalog sheets, price quotations, display cards, and installation instructions.

MERCHANDISING

Merchandising to enhance the convenience and pleasure of purchasing is fast gaining recognition as a major marketing concern and opportunity. As more products are put on the market and as competition between products and between forms of distribution increase at an ever faster rate, merchandising is becoming a critical factor in product success.

The "potato chip theory," built on the premise that you can't eat just one potato chip, also applies to wood products. For example, families who reorganize storage space typically don't stop with one closet. They keep adding on as their needs and income change. Retailers and wood products manufacturers who don't pay attention to such trends are missing merchandising openings.

Using imagination and flair, home improvement retailers are promoting storage systems featuring the advantage of closet, cabinet, and garage storage systems to make a home more comfortable and convenient and add to its resale value. Innovative retailers position themselves as the place to go for storage design information. Storage systems are moved to the front of the store, salespeople stress savings of time and space and are trained to answer questions. Suppliers' displays and merchandising materials help customers design workable systems.

Wood products suppliers who provide retailers with these displays and materials gain an immediate competitive edge. Packaging is an important ingredient in marketing these products, because it helps to create gift-giving opportunities. Packaging also adds convenience to an ordinary product.

More and more home improvement centers are providing systems design and installation services and the Manufacturer's installation instructions can help position its products for the buy-it-installed market.

Establishments often reflect tradition and custom and are slow to change. For example, banks are conservative institutions and bank buildings and interiors reflect these traditions. Most bank products are like commodity lumber products. Deregulation and competition from AT&T, Sears, American Express and other non-banking businesses have forced the banking industry to rethink the merchandising of services.

Noting that retailers spend billions of dollars a year on store interiors that invite shoppers to browse and linger, banks are slowly converting the reserved and stand-offish atmosphere of vaulted ceilings and cashiers separated by high counters to the look and feel of a living room. A cozy club atmosphere and museum quality antique furnishings invite the customer to bide a while. Many banks that have implemented new design and merchandising techniques have shown gains in deposits and loans.

The forest products industry, like banking, has been tradition bound. However, conditions are now favorable for traditional lumber yards and retail dealers to improve their merchandising as have banks. Home center stores have succeeded in creating a merchandising ambiance that makes buying lumber and rocking chairs an interesting adventure.

Merchandising ambiance comes from activities tailored to the customer's tastes. Soft background music that may attract older customers turns the younger crowd away, while videos may discourage older customers. Alert retailers who want to gain and retain customers would do well to assess their customers' needs before investing their cash.

VIII
PRICING
. . . if the price is right

CHAPTER SUMMARY *Pricing is one of the principal competitive weapons in marketing, although the accepted wisdom today is that few companies can compete solely on price. From the manufacturer's perspective, the common denominator is profit, but some companies also have additional pricing objectives. Most forest products companies use cost-plus rule-of-thumb methods to set prices. Using the common rules-of-thumb may result in prices that are too low for company survival or too high to be competitive. Pricing is secondary to firms that offer differentiated products.*

From the customer's perspective, the right price is the one that gives the most value. Price is secondary when buying premium or specialty products.

The price industrial customers are willing to pay depends on the buyer's role in the purchasing decision and the type of industrial products purchased.

Probably no phrase is uttered more frequently in buying and selling transactions than: "If the price is right." The customer says: "I'll take it, if the price is right."

Meanwhile, the manufacturer is quickly figuring the "right price."

Price is the arena where marketing and economics meet. It is the final test of a marketing strategy. If a company plans to market its products on the basis of its differential advantages rather than on low prices, and the differentiation does not attract customers, it may be forced to lower prices. Price defines the value of an item.

No decision is more important in marketing than pricing. Opinions on the importance of pricing as compared to the other three marketing factors—product, distribution, and promotion—have varied over the years. One widely cited study found that back in 1964, business executives ranked pricing as the third most important marketing factor, behind product and promotion. By 1979 executives had moved pricing up to first place. But in 1990, according to an article in the May 1990 issue of *Inc.* magazine, executives said the accepted wisdom these days is that you can no longer compete solely on price. Customers need more service, speedy delivery, and marketing support. Competing on price is passe. It's fair to say price is now probably the least effective arrow in your marketing quiver.

The *Inc.* conclusion appears to be substantiated by Bush, Sinclair, and Araman of the Virginia Polytechnic Institute and State University in Blacksburg, Virginia, who surveyed hardwood lumber buyers in the furniture, cabinet, dimension, flooring, and molding/millwork industries on their satisfaction with the lumber they purchased. Ninety-one percent of the buyers reported that they were dissatisfied with the hardwood lumber they purchased. Deficiencies in lumber quality dissatisfied 61 percent of the buyers most, while price was a source of only 15 percent of the dissatisfaction! The most important supplier characteristic, according to the buyers, was the ability to provide firm price quotations—not lower prices. Can this mean that lumber producers may underestimate the importance of quality and overestimate the importance of price?

Pricing is a competitive weapon, but only one of the weapons available to manufacturers. General Motors doesn't try to compete on price alone; neither do local department stores, gift shops, or businesses selling premium products.

An exception is retailers who promise to match the lowest advertised price a shopper can find in competitive stores. Retailers claim that price-matching promises respond to the fact that consumers compare prices more often now than previously, particularly with major items. Though few shoppers bother to collect on guaranteed matching prices, advertising price-matching does work. Price matching makes products and stores more attractive to buyers.

From both the producer's and consumer's perspective, pricing is one of the most flexible variables in marketing. The producer can change prices, if necessary, and the consumer can choose between product features and product prices in making a purchasing decision.

PRICING FROM THE PRODUCER'S PERSPECTIVE

Pricing Strategies

A company has several choices in establishing a pricing policy: It can attempt to be a cost leader, it can attempt to improve its competitive position by offering differentiated products, or it can combine competitive prices and differentiated products in a strategy aimed at a particular segment of the market.

Companies that aim at cost leadership usually concentrate on competitive pricing, rapid delivery, efficient procurement of raw materials, and employment of trained and experienced personnel. They invest in new processing equipment to increase productivity and decrease processing costs.

If a company does adopt a strategy to compete by selling at the lowest possible price, all decisions must be made to produce at the lowest possible cost. These decisions include buying in bulk, low cost promotion, selecting a market segment, meticulous cost accounting, and rigorous control of all frills.

> Jan Bell Marketing, Inc. sells $181 million worth of private label jewelry to wholesale clubs throughout the country at prices about one-third less than its competitors, and it makes higher net profits than the industry average. Buying in bulk, paying cash for all purchases (earning appreciable discounts), volume production, keeping overhead to a minimum, contracting out most of its assembly work, shipping out production immediately (to save warehousing and inventory costs) all enable Jan Bell to sell at low cost and make a good profit.

Pricing is secondary to firms who improve their competitive position by offering differentiated products. Robert Bush and Steven A. Sinclair, Professors at Virginia Polytechnic Institute, found that companies who have adopted a differentiation strategy place their efforts in:

- ❑ Market research
- ❑ Using new marketing techniques or methods
- ❑ Developing brand identification
- ❑ Serving special geographic markets
- ❑ Promotion and advertising
- ❑ Serving particular customer groups
- ❑ Ability to manufacture specialty products
- ❑ Developing new products
- ❑ Maintaining a company sales force
- ❑ Controlling channels of distribution
- ❑ Owning timberlands or logging operations (considered least important)

Smaller companies that lack the financial resources to be cost leaders succeed by offering products with differentiated advantages to a specific market segment. They focus on what they can do: customer service, rigorous quality control, developing a good reputation within the industry, and operating their production facilities as efficiently as possible.

Prices and Profits

From where the manufacturer and distributor sit, what to charge is uppermost. If the manufacturer doesn't charge enough for the product, the company is not likely to survive. If it charges more than customers want to pay, or more than competitors, the company may price itself out of business.

Survival is the basic pricing objective shared by all companies. Prices may be adjusted to obtain working capital for the operation in economic downturns. In retailing, prices may be adjusted to build traffic.

Although profits are the common denominator of pricing, pricing objectives vary from company to company. Some businesses aim to get the most profit they can without upsetting the delicate balance between increased revenue and the increased cost of earning it. Other businesses are satisfied to receive what they consider an acceptable rate of return on sales.

Good price decisions are based on updated knowledge of customer segments and competition. They also depend on accounting practices that accurately reflect costs.

Almost every company uses pricing in some way or another in an attempt to improve profits:

Larger companies often set pricing based on a percentage of sales or investment. The most commonly reported pricing objective is a specified rate of return on investment. Financial statements show return on investment (ROI) as, for example, profits reaching 15 percent or 20 percent ROI. In Figure VIII-1, an ROI is calculated at 10 percent. If the company sets its target return objective at 15 percent return on investment, the net sales must be increased, or costs decreased to achieve this return. Prices are increased to reflect the desired ROI.

Other businesses base prices on a market share objective to maintain or increase the share of the market. They set the market price below average, if the goal is to gain market share, temporarily sacrificing gross margin. It is common to lower prices to win orders from specific high volume customers if the competition cannot or does not lower prices. This is effective for a new product in a growth market. Market share and profitability have been shown to be closely related for larger companies. One analysis showed that companies with 10 percent to 20 percent market share had pretax profits of about 3.4 percent as a percent of sales; but those with over 40 percent market share had 13.2 percent pretax profit.

A new pricing tool just moving into manufacturing is yield management, which alters prices as the demand for a product changes. This is a pricing technique applied by the airlines to attempt to even out profits between highly used routes and lesser used routes. Yield management is said to be most useful for industries with heavy fixed costs. Utilization of yield management as a price setting technique depends on a reliable

FIGURE VIII-1
Target Return Pricing

1. Net sales		$100,000
2. Cost of goods sold		60,000
3. Gross margin		40,000
4. Sales and general expenses		
Sales	5,000	
General and administrative	20,000	25,000
5. Operating profit		15,000
6. Cost of borrowed capital		5,000
7. Earnings before taxes		10,000
8. Income tax		5,000
9. Earning after taxes		5,000
Capital borrowed		25,000
Equity		25,000
Total		$50,000
Return on Investment (ROI) 5,000 (Earning after taxes) =	10%	
50,000 (Total Capital)		

and complete data base for each product, and sophisticated software. It is not for every company.

These concepts often appear irrelevant to smaller companies. They don't have the resources to fight for an appreciable share of the market. Smaller companies often have a difficult enough task generating profits without trying to also accomplish other objectives like penetrating new markets, or capturing market share, or discouraging new entrants into the industry.

Maintaining market leadership through the product lifecycle is the major objective of some companies. Price decisions during product development influence manufacturing, distribution, promotion and other marketing factors.

> Pricing decisions are adapted to the product lifecycle as the product is developed and goes through the four lifecycle stages—introduction, growth, maturity, and decline. As the product goes through these stages, pricing policies change as follows:

- Introduction stage

A company may adopt a skimming policy on an innovative product, setting a high price on the new product in the hope of recovering development costs and making a profit.

In introducing a new product that competes with others with similar characteristics and function, the introductory price may be set low to penetrate the market and gain market share.

- Growth stage

As sales volume expands, price reductions may be considered as the company gains experience and is able to decrease production costs. Pricing at this stage serves to establish market position and combat competition.

Trade discounts may be used to strengthen relations with dealers.

- Maturity stage

As the product reaches maturity, prices may vary more than in any other stage. Products may be priced defensively against competition. This may be the time to increase options, quality, and services with a commensurate increase in prices to hold or gain new customers.

Or prices may be lowered to fight competition. Companies sometimes attempt to control the market by setting prices low to hasten the exit of marginal firms (reducing competition) and maintain a price leadership position.

- Decline stage

After the product has reached the decline stage, market share is secondary. Prices are set to maintain profits. Prices may be increased for loyal customers to cover costs; or decreased to close out the product.

Some companies concentrate on maintaining stable prices, preferring to focus their energies on product improvement or better promotion. As consumer activism increases, many companies hope that stabilization of prices will win for them a customer perception of fairness.

An exception to pricing to maintain profits is the pricing of upscale items. With these products, prices have little to do with profitability or sales volume and are set relatively high to enhance a quality image. Image is an essential ingredient of most high-priced-quality products, including those made of wood. Should Neiman-Marcus or Bloomingdale's lower prices to bargain basement levels, their quality images would suffer. Another example of how price supports a premium image is Heinz Company, a leader in producing catsup. Heinz keeps its prices above the lower quality competition to support its public commitment to product quality.

Consumers frequently associate quality and status with a higher price. An example often cited in marketing is one of the company that introduced a well-known Scotch whisky as a low price brand. The low price did not attract customers. The producer subsequently designed a new package for the same whisky, devised a new quality image brand name, increased advertising, and doubled the price. Customers happily buy it today at double the price for the status and prestige it brings.

Price, advertising expenditures, and profits are closely intertwined. Analyses have shown that the higher the advertising expenditures of a company compared to the average selling price of a product, the greater the ROI.

Cost-Plus Pricing

In practice, few smaller companies engage in the mathematical computations necessary for target return objectives, market share, or similar techniques. Cost-plus pricing is the method forest products companies commonly use for setting prices. The cost-plus mechanism may be a simple addition of a markup to costs, or a sophisticated pricing formula using a computer. Cost-plus pricing is commonly managed by a rule-of-thumb in which costs are marked up by some formula to determine selling prices. These rules-of-thumb are used so widely that they have become business gospel.

> Four rules-of-thumb are commonly used by manufacturers and distributors to determine selling prices of products:
>
> 1. Selling price is set by multiplying all direct costs by two.
> 2. Selling price is calculated by adding 40 percent of direct costs to the total direct costs of labor and materials.
> 3. Selling price equals three times the cost of labor plus two times the cost of materials.
> 4. Selling price is the direct cost of production plus overhead and profits.
>
> Using these rules-of-thumb, assume the following:
>
> A company sells 10,000 units of an item. At a selling price of $10 each, the company grosses $100,000 in sales and generates a profit (at 10 percent ROI) of $5,000. The cost of material is $20,000. The cost of labor is $40,000. The direct costs of labor and material total $60,000. Overhead costs are $25,000.
>
> Using these costs, the rules-of-thumb calculate four different prices.
>
> 1. Direct costs of $60,000 multiplied by two equals an income of $120,000, which, when divided by 10,000 units, gives a unit selling price of $12.
> 2. Direct costs of $60,000 plus 40 percent equals income of $85,000. Unit selling price is $8.50.
> 3. Labor costs of $40,000 times three and material costs of $20,000 times two equals an income of $160,000. Unit selling price is $16.00.
> 4. Direct costs of $60,000 plus $35,000 in overhead (including the cost of borrowed capital and income tax), plus $5,000 profit. Unit selling price is $10.00.

Using Rule 2, the unit selling price of $8.50 would not earn sufficient revenue to repay the cost of borrowed capital and income tax and still yield a 10 percent profit.

The Rule 3 selling price of $16.00 is probably not competitive. The $12 selling price suggested by Rule 1 looks like the gravy train for the company, providing competitors aren't charging less.

A major problem with pricing based on costs is that so many companies are unable to obtain accurate information on their operating costs. The profit/loss statement may suggest that a business is profitable, but the bank balance shows that it has little cash. In a wood products business, ignoring waste and spoilage is frequently responsible for highly inaccurate costs. An accountant may arbitrarily allocate the difference between the wood actually used in the product and the wood pieces purchased to inventory. The books may be balanced, but they don't accurately reflect the costs of the products.

The rule-of-thumb a company adopts rests to a large extent on how accurately it accounts for costs. A smaller firm with one office-person whose duties include answering the phone, typing invoices and letters, keeping the books, picking up the mail, and generally acting as a gofer, rarely can afford detailed accounting. Every company is required by law to maintain payroll costs, and material costs are available from supplier, utility and other bills paid monthly, therefore Rules 1, 2, and 3 are frequently used.

Unfortunately for the manufacturer or wholesaler, prices are not set in a vacuum. The prices that are best for company profits solve only one of the problems in pricing. Competition, company resources and practices, custom, tradition, and habit all play major roles. Selling-oriented firms are forced to set prices to meet or follow the competition, with little regard for the impact on company profitability. In marketing, the function of cost in setting prices is to determine how pricing alternatives affect profits.

Cost-oriented pricing does not account for variations in product demand. Rigid procedures for developing prices fail to include marketing factors of competition, stock turnover, mark-ups, and other pricing tools. Economic conditions also affect pricing. Lumber prices fall with falling demand, irrespective of the costs of logs or labor. High interest rates, foreign competition, rising labor or energy costs, and similar economic variables override other pricing considerations.

When Sales Decline

When sales decline, a company may respond in many ways. It may decrease prices, or try to round out the product line, reduce operating costs, or move from commodity type products to value-added, or to manufacturing products with a differential advantage, or to developing niche products.

Reducing prices is one of the techniques many companies use to increase sales. When business is bad, they lower prices in the hope of avoiding disaster. When business is good, they often lower prices in the hope of capturing a larger share of the market. But, say some financial experts who have experience in turning around financially troubled companies, cutting prices often makes matters worse. Cash income from higher margins is sacrificed when business is good, and lower prices often mask the real problem when business is bad.

Cutting prices may not solve a marketing problem. Sales must be increased to maintain gross profits when prices are decreased. Figure VIII-2 shows the increase in sales necessary to maintain gross profits

FIGURE VIII-2

Change in Unit Sales Required to Maintain Gross Profit

Price Change		Change in Unit Sales Volume to Compensate for Price Change				
		Decrease in Unit Sales to Maintain Gross Profit				
Price **Increase**		If Gross Margin is:				
		30%	40%	50%	60%	70%
	+15%	–32%	–25%	–21%	–19%	–18%
	+10%	–26%	–20%	17%	–12%	–11%
	+5%	–12%	–11%	–9%	–7%	–6%
		Increase in Unit Sales to Maintain Gross Profit				
Price **Decrease**		If Gross Margin is:				
		30%	40%	50%	60%	70%
	–15%	+100%	+60%	+43%	+35%	+30%
	–10%	+50%	+32%	+27%	+21%	+19%
	–5%	+20%	+13%	+11%	+10%	+9%

Example: If the Gross Margin is 40% and sales prices are increased 15% unit sales can decrease by 25% and still retain the Profit Margin.

If, however, prices are decreased by 15%, unit sales will have to be increased by 60% to maintain the 40% Gross Profit Margin.

when prices are lowered. Moreover, a price decrease often confuses the product image and positioning. Selling a Cadillac at Chevrolet prices results in customers who are not sure what they are paying for. Moreover, cutting prices often tempts salespeople to become mere order takers.

Some companies attempt to increase sales by rounding-out a product line and offering new products at a lower cost. Sometimes this maneuver turns into a trap.

> A small company manufactured a well-known, profit making potting soil for plants. To round-out the product line the company added dehydrated cow manure, decorative stones, and other companion items. Forced to charge low prices for these additional products to meet competition, the company lost money. Its real problem was that it was unfamiliar with the technology of the other products and couldn't afford to produce them efficiently.

Before considering a price cut, other marketing factors ought to be considered. Perhaps the positioning of the product needs to be changed, or the product itself needs a strengthening of differential advantages. Perhaps a new marketing segment should be sought. If nothing else works, a price change may be all that's left.

Failing all else, most organizations will reduce operating costs by downsizing their operation (from two shifts to one shift, or closing a production line). They will endure short term losses, fire the manager or CEO, or take other measures just to continue operating. If the firm can earn enough to pay taxes, mortgages, wages, and other bare essentials it will continue to operate during a downturn. Survival becomes the fundamental objective.

By moving away from the traditional commodity cost and production orientation to product differentiation, many companies in the forest products industry are better able to survive and compete in a business environment in which competition has become increasingly fierce. Adding value to lumber production through kiln drying and planing has caught the attention of the forest industry. The desire to lessen dependence on the construction industry and penetrate other industries has spurred the move to further processing to produce differentiated or niche products for a diverse market base. Price changes affect market share, positioning of the product versus competitive products, and relationships with distribution channels. A relatively small price change can have a large impact on the bottom line of a profit/loss statement.

Other common situations may prompt a company to consider changes in price:

- ❑ Prices may be too high or too low compared to competitive prices for the same product.

- ❑ Customers may object to price differentials among items in a product line, or may not understand why the prices are different.

  Beatrice Foods Co. sells all of its Dannon yogurt flavors at the same price, even though some yogurts cost the company more to make. This is called price lining. Price lining is based on the principle that when the confusion of different prices for the same product is eliminated, people are encouraged to buy.

- ❑ Prices set low to attract customers may be wasted on the type of customer who has no brand or product loyalty.
- ❑ Public opinion may mobilize against prices, if they feel that a company is exploiting a situation. During the Middle East crisis of 1990 gasoline prices rose sharply in many areas, resulting in strong public opinion against the American oil companies.
- ❑ A change in a customer's purchasing policy may force reconsideration of prices. Figure VIII-3 narrates a dilemna this situation poses for a company.

FIGURE VIII-3
Handling a Pricing Predicament

A predicament that producers of industrial products find themselves in at one time or another stems from customers who change their purchasing policy in an aggressive attempt to curtail costs of production. The supplier who has heretofore comfortably priced products finds himself competing with cost cutting foreign rivals and hungry competition.

The following problem is a case study of the type developed at the Harvard Business School and reported in the *Harvard Business Review* of March-April 1988. The pricing predicament is adapted to the forest products industry. The seller is a producer of dimension components with a reputation for producing quality products and a long standing policy of not cutting quoted prices. The buyer is a furniture manufacturer who is under the gun to lower the production costs of one of its products which has been copied by the competition.

The buyer invites bids for a hardwood component for its standard furniture line. Three bids are submitted: one at $43,000 from the dimension components producer with the fixed price predicament, who has supplied components to the buyer for years; another at $40,000 from a supplier from whom the furniture company has never purchased before but who has a good reputation; a third at $39,000 from a new domestic mill.

(Continued)

FIGURE VIII-3
Handling a Pricing Predicament (Continued)

The $39,000 offer includes no quality warranty and is discarded. The purchaser informs its usual supplier that they will place this order with the competitor at $40,000. They also say that they would prefer to buy from their usual source if the supplier will reduce its price. The supplier's salesman unsuccessfully attempts to rescue the order by pointing out the weaknesses of the competitor's service policy.

The purchaser is an expanding company and there are many desirable future orders at stake. The supplier's salesman recommends to his company that they rebid the order at $42,000, which he believes the purchaser would pay for the security of dealing with someone they know. The dimension producer rejects the idea of setting a precedent for all orders by rebidding at a lower price.

The salesman recommends that his company consider three optional actions to rescue the order and still maintain its fixed price policy:

❑ Trim back the usual on-site service provided the buyer

❑ Eliminate the zero-defects warranty and the JIT delivery policy for this order

❑ Some combination of the first two alternatives that would lower costs

The situation poses a dilemma for both buyer and seller. The buyer has to lower costs, yet wants to deal with a company on which it knows it can rely. The seller wants the order but does not want to break the fixed price policy which has helped it earn a reputation for quality.

The supplier now is faced with considering a change in policy or losing a valued order.

Four business executives consulted on this problem offer four different approaches to the dimension supplier's pricing problem:

❑ Approach #1: Don't concede on price. Solve the problem by getting closer to the customer and selling the customer.

The fixed price policy tells the supplier's sales organization that the company expects them to know their customers so well that they understand their customer's business and strategies more thoroughly than any other supplier.

The supplier should provide such a high level of service to its customer that they develop a working partnership to solve the customer's problems. The fixed price policy also indicates that a pricing decision shouldn't be made by the sales department alone. The production and service departments should have input on how the price is determined and the other factors that go into the sale. For instance, the component is for a new furniture line. It might be advisable for the supplier to offer a short run at the unit bid price for the buyer to try out. The buyer and seller could work together to determine if the component meets the buyer's

FIGURE VIII-3
Handling a Pricing Predicament (Continued)

requirements. The objective is to keep a customer, rather than make an individual sale.

❑ Approach #2: Negotiate or lose the business.

The buyer is looking for greater perceived value. One way to accomplish this may be to unbundle the bid, that is, set a separate price for the product itself, and price the zero-defects warranty and JIT delivery separately. In that way, the service provided by the usual supplier and the competitor can be compared to assure that the comparisons are equal. Reducing service might bring the bid down to the competitive price and still assure the customer that he is getting greater value from his usual supplier. This may not maintain the dimension supplier's usual profit margin, but the company may be able to make up the lower price by cost reductions in manufacture.

❑ Approach #3: Price is not the real issue. The real issue is value to the buyer.

Value to the buyer is measured in terms of the effect of the component on the success of the buyer in gaining its competition edge. The relationships between industrial suppliers and purchasers are critical in sustaining competitive performance for both parties. The component has to help both the buyer and supplier get and keep a competitive edge.

Price is only one factor in competition. It ranks fourth in furniture buying. The salesman should be working on helping the customer meet its competitive goals. Production, engineering, and the supplier's purchasing department should be in on the discussion on how to supply value to the customer. For example, the supplier might suggest using a different species, a different glue, a slight modification in design might accomplish the objectives of both buyer and seller.

❑ Approach #4: The supplier's sales organization has only itself to blame.

The supplier's sales department had been notified earlier that the buyer had to cut production costs to meet its competition and was going out for competitive bids. But the supplier had not taken the warning seriously. The supplier's high regard for its reputation for quality and reliability had developed a false sense of security and a feeling that no competitor could pose a threat to its sales. The sales department is arrogant if it does not recognize the advantage the buyer has offered in being willing to pay more than the lowest bid. This dilemma is not the salesman's problem; it is the supplier's problem. Management, production, design, and other facets of the company have to be part of the solution for both the customer and the supplier.

A comment from the buyer's viewpoint was offered by Dave Groom, Senior Project Engineer of Pennsylvania House, Lewisburg,

(Continued)

FIGURE VIII-3
Handling a Pricing Predicament (Continued)

Pennsylvania, who reviewed this case history. Groom has been responsible for the purchase of many thousands of dollars worth of dimension components. He minimizes the risk of dealing with a new supplier by "testing" with 10 to 20 percent of the order to start. The supplier facing the pricing predicament in this case might continue to service the customer by checking to determine if a test order and subsequent orders met the buyer's expectations.

THE CUSTOMER'S EYE VIEW OF PRICE

The narrow view of setting prices based on production costs, cash flow, and profit requirements is changing. Firms are now adopting the practice of buyer based pricing—using pricing as a tool for creative marketing. Prices are set in the design stage, based on the purchasing motivations of the buyer.

Attention to how buyers value a product, who buyers are, why they buy, and how they make purchase decisions is as important to determining the right price as it is to determining the right product.

Buyers value the same product benefits differently. In modern marketing, from the customer's perspective, the right price depends on:

- ❑ How much the buyer needs the product
- ❑ How much the buyer wants the product
- ❑ The value of the product to the buyer
- ❑ The buyer's desire for status or some other emotional gratification

The right price depends also on whether the customer buys as a consumer for his or her own use, or for someone else's account, as in industrial purchasing. The decision to buy for a customer's own use is likely to be subjective. Industrial decisions are more objective.

The price that buyers are willing to pay depends also on the type of product they intend to purchase and what they intend to do with it. More importantly, it depends on their perception of the value of the product to them:

- ❑ When buying food products, health and beauty aids, cleaning supplies, and other consumer packaged convenience goods, the consumer's perception of value is shaped by how the items are packaged and advertised.

- ❑ In purchasing appliances, automobiles, homes, and other consumer durables and shopping products, the customer can understand price differences between similar products by observable differences. Advertisements will have informed the consumer about product features, benefits, and prices. Price is an important factor in the purchase of products which are viewed as essentially the same.
- ❑ Price is usually secondary in the purchase of cameras, gifts, and other specialty goods. The use of the item often influences what the buyer is willing to pay. A rolling butcher block table may command a higher price if it is to be given as a gift than if it is to be used by the buyer. Specialty papers for gift wrapping sell for considerably higher prices than the same papers might sell for the buyer's everyday use.

Shopping products, in which price is the major factor in a buying decision, can be changed to specialty products by using the concepts of singularization and product differentiation. A customer may view and value one hardwood floor the same as another, and pay accordingly. Singularize the ordinary hardwood floor to "patterned oak from Pennsylvania's prime oak country" or differentiate it by making "flexible maple flooring that combines the durability of maple with the resilience of carpet" and the floors acquire the value of specialty products.

The Industrial Customer's View of Pricing

The original quoted price in industrial purchasing may be changed by bidding and negotiation and a system of discounts. The price an industrial customer is willing to pay is influenced by the person's role in the purchasing decision and the type of industrial products purchased:

- ❑ When buying *support consumables*—paper clips, welding gases, cleaning and other supplies—buyers look for low prices. These consumables have list prices, but distributor specials, mark downs, and give aways make these prices more fictional objectives than real prices, as illustrated in Figure VIII-4.
- ❑ In purchasing *process consumables*—lumber, panel boards, dimension, and similar commodities consumed in the manufacturing process—competitive pricing is the custom.
- ❑ Competitive prices are critical in *make-or-buy consumables*—items the purchaser can buy or has the capability to make in-house. Competitive prices are not always the determining factor.

  One survey of furniture manufacturers showed that quality and JIT delivery were major factors in the price buyers were willing to pay. Dependability of the supplier was almost of equal importance.

FIGURE VIII-4
List Prices: Fact or Fancy?

Market and list prices are often unrelated. List prices are often used as a base from which prices are discounted for special classes of customers, while in some product lines a steady percentage of sales are sold at full list price, or near list price.

Analyses have shown that discounts between 10 percent and 40 percent produce little additional sales volume for construction materials and other industrial products. There appears to be an assumption on the part of the sales force that automatic discounts exceeding 40 percent are essential.

It is claimed that buyers purchase industrial equipment more readily at a price discounted 55 percent from list, than they do at the same price, if the price is presented as based on cost.

Companies usually grant full discounts on large orders in which bidding is competitive. Chain or corporate buyers who place repeat orders also receive full discounts.

Small specialty orders that require considerable custom design are often fully discounted, but pricing experts maintain they should not be.

Special products ought to have special prices. List prices detract from the uniqueness of these products.

Products that are usually heavily discounted have a tendency to be viewed as commodity items and a commodity mentality develops about their pricing.

Source: "What Good is List Price?", James C. Makins, Pace Magazine, October 1988, reprinted in Marketing 90/91, published by The Dushkin Publishing Group, Inc. Sluice Dock, Guilford, Connecticut

Price, quality, and delivery are less important than putting existing production facilities to use. When the National Dimension Manufacturers Association studied the factors influencing furniture manufacturers' make-or-buy decisions, they found that many manufacturers had previously invested in plants and equipment to manufacture their own components. These manufacturers felt it was more important to their profit/loss statement to utilize this investment to produce dimension components than to buy out. In this instance, price was forced to compete with other overriding objectives.

- When buying capital goods—expensive items that may require the approval of the CEOs, an engineer, designer, plant manager, or other executive—the purchasing decision may take several months or years of discussion before a purchase authorization is signed. Pricing has to accommodate changes in costs over such long decision making periods.

- ❑ Turnkey pricing is a new trend in setting prices for industrial products.

  The turnkey division started by Anthem Electronics, a manufacturer of industrial electronic components in San Jose, California, provides a model that dimension component producers might adopt. Anthem Electronics turnkey operation takes on the job of supplying parts for a customer's product list, building the component, testing it, and providing the part ready to use.

  Furniture and cabinet manufacturers faced with increased worldwide competition are searching for methods to cut manufacturing costs. A producer of dimension components could provide a similar turnkey service, using the knowledge of where to obtain dimension of different species, and the processing equipment and know-how. A turnkey service adds considerably more value than merely supplying parts. Customers are freed to concentrate on design and marketing, while the dimension component producer concentrates on assembly.

  In Germany, for example, it is common for a cabinet door supplier to develop new lines for cabinet manufacturers and negotiate prices based on the designs selected by the buyer.

Other factors that influence the price a manufacturer charges an industrial purchaser are:

- ❑ Customer demand

  How seriously does the customer want or need the item?

- ❑ Competition

  Can the customer buy the same or similar item at the same or lower price?

- ❑ Value considerations

  What value is the customer receiving for the price paid? The value may be considerably more than the price of the direct labor and materials that go into making the product. Installation, warranties, service, packaging are all values that may be priced into the product.

- ❑ Company pricing objectives

  What does the seller require from the price—increased sales, survival, profits, increased profits, a specified ROI? These determine how much give the quoted price has.

- Company management

  Company marketing strategy decisions play a large part in price negotiation. The firm that adopts a cost leadership strategy concentrates on low prices. The company with a product differentiation strategy focuses on the differential advantages.

- Government regulations

  Federal and state governments have created a regulatory framework that affects pricing. Increased pricing regulations and increased enforcement of legislation restrict the freedom of producers in arriving at prices. Restraint of trade regulations, especially, pose limits on pricing actions.

In 1979, a number of Chicago house builders and their material suppliers offered to do their bit against inflation by adopting a 60-day price freeze. Although the action was intended to be part of the national commitment against inflation, then out of control, the Federal Trade Commission vetoed the action as a possible restraint of trade.

IX

RESEARCHING THE MARKET

. . . discovering markets and customers

CHAPTER SUMMARY *Marketing involves finding markets, understanding buying behavior, and learning about customer's needs and wants. Market research is a method for discovering more about a market at minimum cost and with maximum efficiency. Market research is a five-step process for acquiring data and analyzing the information in making marketing decisions.*

The United States Census is the major source of information about the American people and current markets. Information is also available in publications at libraries. This chapter explains how to use such sources to acquire information about competitors, investigate the market potential for expanding sales, evaluate the market size for a new product, and develop marketing strategies. Reference books for a minimum desktop information library for market research in the wood products industry are suggested.

Great marketing ideas are often conceived at odd times—at 3:00 AM, during the morning shower, or perhaps while taking a walk. Marketing decisions on those great ideas are frequently made during the first flush of enthusiasm. The shoot from the hip method is often successful. But many of these great ideas miss the target entirely, adding another notch in the statistics of business failures. Shooting from the hip is a luxury no one can afford anymore. Mistakes are too expensive; markets too complex. There are too many products competing for customers. Today, marketing a wood product is too risky a proposition to rely on only a manager's gut feelings. Hard data is necessary to confirm ideas.

To minimize risks in marketing, to better assure that the great idea hits the target, companies conduct *market research.*

Market research minimizes the risk in marketing decisions by systematically gathering, recording, and analyzing data to solve marketing problems. Market research is about discovering, understanding, learning, investigating, and evaluating.

Akio Morita, Chairman of Sony, once said: "Cut a thin slice, but cut deep." Market research helps a business determine how thin is "thin," and how deep is "deep."

The answer to these questions is market research. Market research can be done on a shoestring or it may require a sizeable budget. Expenditures for market research vary with the size of the investment in the product, the size of the company, and the objectives of the research. The research may consist of one-on-one interviews with key customers, large-scale mail or telephone surveys, focus-group interviews, or other methods of acquiring information. Each technique serves a specific purpose.

Market research is the foundation for product development, market segmentation, pricing products and services, promotion activities, and marketing strategy decisions. A credible market research study may quicken the interest of an investor in a business venture. It provides an objective base for business planning.

MARKET RESEARCH: GUARANTEE OR PROBLEM SOLVER?

Market research has sometimes been considered a panacea for business ills. Conduct thorough market research, the conventional wisdom says, and you'll produce a winner. That does not always work. Ask the Ford Motor Company about its Edsel car that fizzled in the 1950s or Coca-Cola about its New Coke that fizzled in the 1980s. Both products were well researched and both failed to capture the anticipated market share.

Overlooking a detail may jeopardize an entire conclusion. The failure of the magazine Literary Digest in the 1930s is a classic in market research. The Digest conducted a major national telephone study on voter intentions in the presidential election of 1936 and headlined its conclusion that Alf Landon would beat Franklin D. Rooosevelt. When Roosevelt won a stunning victory, market researchers explained the fiasco by noting that the research technique used failed to realize that in 1936 many voters did not have telephones; therefore, the survey did not represent the majority of voters.

Economic forecasting is often called the "dismal science." The popularity and success of economists in predicting gross trends like the growth of gross national product (GNP), inflation, recessions, unemployment, and population has ebbed and flowed over the years. During the 1950s and the early 1960s, when the U.S. economy had a relatively stable growth, the models used by economists for forecasting trends worked comparatively well. But in the 1970s and 1980s, the neat models based on mathematical equations were subjected to so many exogenous, or unpredictable, factors that actual happenings were often quite different from the economic forecasts.

Macroeconomic forecasting is giving way to microeconomic forecasting, examining individual industries instead of the wiggles in the GNP, just as the old mass marketing is giving way to targeted marketing.

As a method for solving problems, market research has demonstrated significant capability.

> Johnson Wax is a successful marketer of hair rinses, shampoos and other superior personal care products. Much of their success is attributed to thorough market research. The introduction of their Agree hair products utilized market research in every step to identify opportunities, specify the target user, devise the marketing strategy, and determine what potential customers wanted and expected from the product.
>
> Market research helped sort the products to be promoted from the products to be rejected. (Hair dressings, hair coloring, and hair sprays were rejected.) In-depth interviews revealed the product characteristics customers preferred. The result of the careful research was a 20 percent market share achieved for the Agree products within two years.

WHAT A WOOD-BASED BUSINESS CAN LEARN FROM MARKET RESEARCH

Market research is used to answer the kinds of questions business managers ask about marketing, such as:

- ❑ What is our product line's market share?
- ❑ How can we expand sales of these products?
- ❑ What is the market potential for a new product?
- ❑ How many potential customers are there?
- ❑ Are potential customers local, statewide, regional, national, or international?
- ❑ How should we segment the market?
- ❑ Is there a market niche our company can fill?
- ❑ Is the market growing?
- ❑ Is there room in this market for a new product?
- ❑ What are the characteristics of the customers for the product?
- ❑ What benefits do potential customers want from the product or service we're offering?
- ❑ What will motivate them to buy?
- ❑ Will the product or service fit new lifestyles and new trends?
- ❑ How will customers learn about the product or service?

- ❑ What type of promotion program should we use?
- ❑ In what media should we advertise?
- ❑ Where should we locate our manufacturing plant?
- ❑ What's the most effective method of distribution of the product or service for these potential customers?
- ❑ Who are our competitors?
- ❑ What marketing strategies do they use?
- ❑ What prices do they charge?

The necessity for market research in the forest products industry has never been greater.

> A long-time sawmill operator with a substantial base of old-growth Douglas fir inherited from his family and a sawmill built to handle large-diameter, long logs was discussing with an associate whether to attend a meeting on marketing forest products. "I've been selling lumber for a long time," he said wryly, "so they're finally going to tell me what I've been doing."
>
> This mill operator read with fair regularity the publications that forecast whether housing starts would go up or down, the anticipated growth of the GNP, and the possibility of a recession on the horizon. Though he often knew who was buying his lumber, his interest was in the macro market: how much he could sell and at what price. As a manufacturer of commodities, macroeconomic forecasts were his major barometer. The system worked well for him, so long as he had ample timber supplies, and the construction market held up.
>
> However, the younger son of this long time mill operator had a different perspective. He realized that he couldn't depend indefinitely on the availability of the long, large timber. He also followed trends and paid attention to the predicted changes that the aging of the nation's housing stock and the high costs of home construction would push remodeling and repair to the forefront. He also recognized that a retail revolution was underway. He decided that manufacturing wood products would be the way to assure long-term survival of his business.
>
> He constructed a plant to manufacture window components and windows. He conducted market research to learn where he should concentrate his marketing efforts, the potential market for his particular product, and the styles and features potential customers wanted. He read with regularity magazines about home decorating trends, studied what competitors were making and selling, talked with retailers and architects, and kept up-to-date on the demographic changes happening in the country. He worked closely with wholesalers and retailers to meet customer demands. He adjusted to the changing resource and changing market and succeeded in his venture.

The different approaches of father and son did not represent a generation gap. They reflected more the changing face of the forest industry in which market research has become the core of manufacturing wood products and products have become the way to gain a competitive edge.

TECHNIQUES OF MARKET RESEARCH

Over the years, many techniques have been developed for systematically gathering, recording, and analyzing data in market research.

Prior to undertaking a formal market research program, the entrepreneur or owner in a small company, or the marketing manager or marketing vice president in a larger business usually undertakes some informal exploratory research. This informal research may take place over a cup of coffee, or at lunch, or in meetings with management staff. Wholesalers, retailers, customers, and other informed persons assist in mulling over the idea. Too often, the exploratory research is both the beginning and the end of the idea. The idea may be dropped, or the informal discussions may generate sufficient enthusiasm to encourage going ahead with the project, bypassing further investigation.

For example, a manufacturer of lattice may have the idea that sales could be increased if a computerized design service was made available to provide customers with exact bills of materials for their projects, including all the materials on the bill packaged, cut to size, and ready to assemble. He undertakes some exploratory research and decides to proceed with further investigation.

The further investigation is market research.

Market research is a six step process:

1. *Determine the information the research should produce, and the problem to be solved*

Is the problem identifying a target market? Or is it obtaining a better understanding of potential customers? Is it identifying a promising market? Or is it determining current preferences in millwork for home decoration?

2. *Acquire pertinent secondary data*

This refers to broad-based data that has been previously published. For example, to learn the size of an industrial market, consult the United States Census of Manufacturers. To evaluate the impact of the aging population, check issues of *American Demographics.* To discover how to reach lower income consumers with a line of RTA storage cabinets, study *Sales & Marketing Management's* latest issue of the "Survey of Buying Power."

3. *Acquire as much primary data as the project can afford*

The secondary data has provided a broad picture. But if a business expects answers to specific questions, primary research is required. Collection of primary data is apt to dent a budget considerably more than the acquisition of secondary data. It frequently involves printing, mailing, telephone, fax, travel, and other costs.

Collection of primary data is also labor intensive. A professional may be able to squeeze in no more than five or six one-on-one interviews a day. A questionnaire mailed to 1,000 companies may garner a response as small as 5 percent to as large as 40 percent. Design of a questionnaire or interview form that produces the desired information is an art in itself.

4. *Analyze the data acquired*

Data is of little use until it is analyzed and interpreted. Failure to analyze and present the results in a form management can use for marketing decisions results in studies that lie unused on the shelves. These failures are often due to lack of rapport or lack of understanding between managers and market researchers.

When management does not include the results of the research in discussions, or requests only bits of information about parts of problems, or does not allow enough time for research, market researchers do not make the contribution to solving problems of which they are capable. It is particularly irksome to market researchers when management acts on preliminary conclusions and relies more on intuition than on research.

On the other hand, managers often complain that market researchers do not organize their research to solve problems, that they are often more involved with the research techniques than the results, and that they don't present results in language management can understand.

5. *Use the analysis to reach conclusions*

The mutual complaints emphasize the importance of satisfying both the methodology requirements of market researchers and management's demands for definitive information to solve problems.

6. *Utilize the research findings*

For example, research findings that light-colored moulding is the preference in current decorating trends, or that home center store customers need lumber or moulding lengths that fit into the trunks of their cars are of little use if the company does not utilize them. Management may determine that it is not in the company's best interests to go along with a trend, but if it makes that decision, it does so deliberately.

TOOLS FOR ACQUIRING SECONDARY DATA

As in all sciences, the value of market research findings depends on both the availability of tools and the skill with which these tools are used. Because primary research requires so much time and cost, small businesses may use staff to collect secondary data and then hire professionals for analysis and acquisition of primary data.

Secondary data is available in almost overwhelming detail. Marketing skill lies in knowing which data to access and having ability to judge the accuracy and usefulness of the data. Sources of secondary data are available in most libraries—free from government sources, and for fees from private sources. The United States Census cannot be equaled as a basic source for secondary data.

THE UNITED STATES CENSUS

The 1990 Census is a remarkable document containing more information on Americans than any previous census. The Census Bureau collects information that no other organization can afford to acquire. The questionnaire used for the 1990 census cost taxpayers $26 million just to print the 100 million mailing packages.

Although the census can be mined for a staggering amount of information, the data has limitations. The Census of Manufacturers is conducted every five years, so by the time it is in print it is out of date. The Census of Population is conducted only once every ten years. Seeing into the future by extrapolating this old data can lead to conclusions that are off the mark. Its terminology can lead a researcher astray.

Nevertheless, the 1990 census provides information not previously available. It includes data on pension income, disabilities, and congregate housing—a boon to those marketing products or services targeted at the elderly. Suppliers of products to home builders and home owners will now find in the census data on home equity loans fees paid by owners of condominiums and mobile homes and how long it takes people to travel to work. The 1990 census provides a better look at affluent homeowners than any previous census. It tells marketers who has solar heating, and how a person's education level relates to spending habits.

The 1990 census boasts a new feature called topologically integrated geographic encoding and referencing system (TIGER) which can help in developing computer maps to plan sales territories, pinpointing direct marketing prospects, or link a geographic area to age, income, or race.

By contacting the Census Bureau in Washington, D.C., a market researcher can obtain a CD ROM based database of census data for the entire United States for less than $10,000. But the numbers are meaningless without specialized, privately produced software. The flood of information available from the census has generated sophisticated desktop marketing systems that allow managers to manipulate, analyze, and interpret data on a marketer's personal computer. This sounds expensive and time consuming and it can be. PC based desktop systems may cost thousands of dollars. American Demographics publishes an annual directory of the "Best 100 Marketing Information Companies." The marketing information field is growing so rapidly that keeping current is a chore in itself. Many of the trade journals publish information for market researchers on software programs currently available.

Marketing information companies range from full service providers, to specialists in psychographics, ethnic marketing, buying behavior, media, direct marketing, and international marketing. Of particular relevance to the forest products industry is NPA Data Services, Washington, D.C. which provides, for a fee, programs or data on markets for construction, and analyses of demand for construction in U.S. metropolitan areas.

A mountain of census data is available from private companies that have developed individual formats. The information is available in paper reports, on diskettes for personal computers, magnetic tapes that can be customized to meet individual marketing requirements, and books.

These expensive systems may intimidate a small business that only occasionally requires data. For these businesses the U.S. Census Bureau in Washigton, D.C. can provide an inexpensive, simple, printed report that may be used to help identify market areas or define new markets.

Census Data for Wood-Based Industrial Products

For manufacturers and wholesalers of industrial wood-based products, the Census of Manufacturers is the mother lode. Based on the *standard industrial classification (SIC),* a system developed by the U. S. government for use in collecting data about industries, the Census of Manufacturers is updated every five years. The reports cover about 450 classifications of manufacturing industries, including wood-based industries.

The organization of the census SIC system for major elements of the forest products industry is shown in Figure IX-1. The SIC codes are divided into ten two-digit broad industry groups. For example, SIC 01–09 covers the agriculture industry, SIC 40–49 is transportation, and SIC 60–67 covers finance, insurance, and real estate. The SIC of most interest to manufacturers of industrial products are codes 20–39.

Manufacturers of industrial wood products are interested in SIC 24, lumber and wood products. SIC 24 includes primary and secondary wood products, ranging from lumber to wood bowls.

The five year reports of the Census of Manufacturers include:

- ❑ Final area reports—statistics on value added by manufacturing, employment, payrolls, new capital expenditures, and the number of establishments for each SIC.
- ❑ Final industry reports—statistics presented in separate reports for each SIC code industry covering value of shipments, capital expenditures, value added by manufacturing, cost of materials used, and employment. The data is cross-classified by geographic region, state, employment size, class of establishments, and degree of product specialization.

Census Information on Consumers

To understand consumers, the market researcher has available other census data. The Census of Population, issued every ten years, is presented in four series:

- ❑ Series A gives the number of inhabitants
- ❑ Series B describes general population characteristics
- ❑ Series C describes general social and economic characteristics
- ❑ Series D cross-classifies Series C by age, race and other characteristics

The four series describe the demographics of the United States. From these, market research experts can piece together characteristics, such as age, sex, income level, and residence of potential customers. Characteristics may be further explored through the Census of Retail Trade, The Census of Wholesale Trade, The Census of Selected Services, and the Census of Housing.

These explorations require enormous expenditures of time and personnel. However, professionals who have invested the time have culled the material and prepared books and special reports on different facets of the market.

FIGURE IX-1

Organization of the Census Standard Industrial Classification (SIC) System

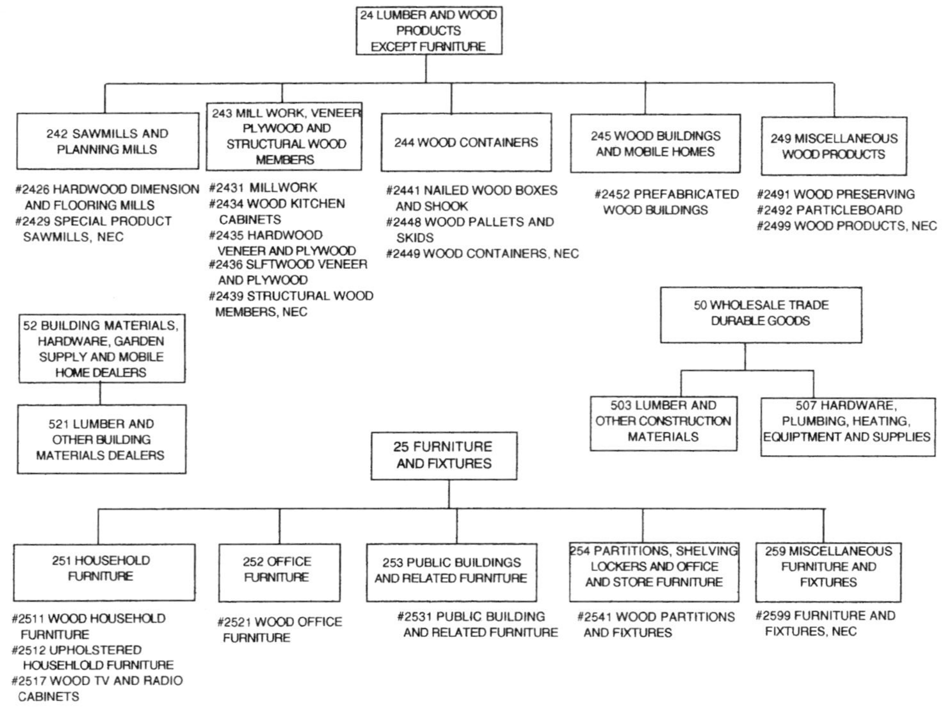

American Demographics summarizes demographic changes, using both census and updated information, in bite-size, easy-to-understand articles. The January 1991 issue, for example, provided demographic information on baby boomers. Other issues have covered trends and opportunities for furniture manufacturers, reaching the elderly market, and marketing children's toys to grandparents.

OTHER PUBLICATIONS

Although the census is a market research gold mine, other government publications provide information of value in research. Figure IX-2 includes the names and addresses of other government publications. Figure IX-3 includes business guides for market information.

FIGURE IX-2
Government Sources of Market Information

Most of the following publications are available in major public libraries and the libraries of universities.

"County and City Data Book"—an annual supplement to the "Statistical Abstract" provides almost 200 statistical areas, unincorporated places, and urbanized areas. Available from the Bureau of the Census.

"County Business Patterns"—an annual volume covering county, state and U.S. summary statistics on employment, number and employment-size of reporting business units, and taxable payrolls for approximately 15 broad industry categories. Data is particularly suited to analyzing market potential, establishing sales quotas, and locating facilities. Available from the Bureau of the Census.

"Economic Indicators"—a monthly digest of current information on prices, wages, production, business activity, purchasing power, credit, money, and federal finance presented in charts and tables. Available from the Superintendent of Documents, Government Printing Office.

"Federal Reserve Bulletin"—a monthly compilation of statistics on banking, deposits, loans and investments, money market rates, securities prices, industrial production, flows of funds, and various other areas of finance in relation to government, business, real estate, and the consumer. Available from the Board of Governors of the Federal Reserve System. Each of the 12 Federal Reserve Districts also has its own publication covering statistics within the district.

"Monthly Labor Review"—a monthly compilation of trends and information on employment, wages, weekly working hours, collective agreement, industrial accidents, and other current labor statistics. Available from the Bureau of Labor Statistics, U.S. Department of Labor, Washington, D.C.

(Continued)

FIGURE IX-2
Government Sources of Market Information (Continued)

"Statistical Abstract of the United States"—an annual summary of statistics on the social, political, and economic organization of the United States. Available from the Bureau of the Census.

"Statistical Yearbook"—an annual publication on international statistics of population, agriculture, mining, manufacturing, finance, trade, and education. Available from the United Nations, New York.

"Monthly Bulletin of Statistics"—a monthly supplement to the "Statistical Yearbook." Available from the United Nations.

"Survey of Current Business"—a monthly publication on gross national product, national income, and the international balance of payments. Each issue updates statistics on general business indicators; commodity prices; construction and real estate; domestic trade; labor force; employment and earnings; finance; foreign trade of the United States; transportation and communications; and several raw material industries. Available from the Bureau of Economic Analysis, U.S. Department of Commerce.

"U.S. Industrial Outlook"—an annual publication of the U.S. Department of Commerce. Available from the Superintendent of Documents, Government Printing Office, Washing,ton, D.C. 20402. The "Outlook" covers industries according to SIC classifications and includes descriptions of industries, trends, employment statistics, and historic data.

Trade publications serving the forest industry compile annual data in special articles, factbooks, or special issues. Business magazines for the appliance industry, supermarkets, computers, discount stores, distribution, drug and cosmetics industry, agricultural implements, men's wear, brewery, petroleum, and other industries also publish special statistical issues. Consulting these special issues may shed additional light on a market. A manufacturer may have an idea for an elegant, small wood box for packaging gift sets of shaving lotion, after shave talc, and related items. The box, which would have a long utilization period, would promote the sale of the shaving items. To discover the size of that market, the researcher would consult a trade journal such as *Drug and Cosmetic Industry's* annual "Drug and Cosmetic Catalog."

To locate the trade publications for an industry, consult *Ayer Directory of Publications* issued annually by Ayer Press, Philadelphia and available in most libraries.

Sales & Marketing Management (S&MM) (Bill Communications Inc., New York City) publishes an annual "Survey of Buying Power," which has become a prime non-government source of data for sales planning,

FIGURE IX-3
Some Business Guides for Market Information

Most of these Guides are available in public or university libraries.

Standard & Poor's Register of Corporations, Directors and Executives—published annually by Standard & Poor's Corporation, New York, in three volumes. Volume 1 is an alphabetical listing of nationally known corporations, including top executives, directors, principals, and annual sales. Volume 2 is an alphabetical listing of directors and executives in the United States and Canada. Volume 3 is a list of corporations by SIC code and geographic area, including individuals and new companies.

Standard & Poor's Corporation Services—several subscription publications are available from Standard & Poor's Corporation, New York. These include *Industry Surveys,* an annual survey of specific industries. *The Outlook* is a weekly stock market letter. *Stock Guide* is a monthly summary of investment data on over 5,000 common and preferred stocks. *Trade and Securities* is a monthly listing of statistics on business and finance.

"Directory of Corporate Affiliations"—an annual publication of the National Register Publishing Company, Skokie, Illinois, lists approximately 3,000 parent companies with their divisions, subsidiaries, and affiliates, as well as an index of who owns whom.

"Moody's Industrial Manual"—published annually by Moody's Investor Service, New York; provides background, business and products and descriptions, history, mergers and acquisitions, principal plants, and properties, principal officers and directors, and financial statements and statistical records for listed companies.

"Standard Rate & Data Service Publications"—a series of publications from Standard Rate & Data Service, Inc., Skokie, Illinois used in preparing and placing advertising in various media. The directories "Newspaper Rates and Data," "Spot Radio Rates and Data," and "Spot Television Rates and Data" also include current statistical market data for states, counties, cities and metropolitan areas organized under population, households, income, retail, and store-type sales.

"Thomas Register of American Manufacturers and Thomas Register Catalog File"—a multi-volume publication issued annually. Includes manufacturers classified by products and services. Also alphabetically lists 60,000 brand or trade names. Available from Thomas Publishing Company, New York, N.Y.

marketing, advertising, forecasting and long range planning, estimating sales potential for specific product lines, and selecting media for advertising. It presents data in a form useful for measuring market potential, developing marketing and distribution strategies, and evaluating the potential of new products. Statistical information to guide the selection and allocation of media and setting up sales territories can short cut experimentation and result in time and cost savings in marketing. Statistical data is presented by product lines such as furniture/home furnishings/appliance store sales and building material and hardware store sales.

Using S&MM statistical information, marketers can determine the percent of the population by age group (important in sports, toys, furniture, and do-it-yourself products), the number of households (for home remodeling products), furniture and furnishings sales, effective buying income, the type of retail stores where products are sold, and the media used. These statistics are cross-referenced by national, regional, metropolitan statistical areas, states, and county city areas.

For in-depth research on customers, a number of business guides are available in many libraries.

Other sources of information are publications of trade and professional organizations, annual reports of companies in the forest industries, and trade shows.

TOOLS FOR ACQUIRING PRIMARY DATA

Sooner or later, a marketer arrives at a point where answers to the questions that prompted the project cannot be found in secondary sources. This is the time for primary research. Acquiring primary data is expensive and requires knowledge and skill.

The basic methods used to acquire primary data are:

- ❑ Observation
- ❑ Surveys
- ❑ Experimentation
- ❑ Simulation

Observation

Visits to a trade show and watching buyers at a retail store are both methods of observation. The results are often not statistically significant, but an astute observer can learn much about buyer behavior from observation.

Surveys

Surveys are the most widely used method of acquiring primary research data. Surveys may be conducted by telephone, mail, or personal interviews.

More than half of all primary research is conducted via telephone interviews. They are relatively inexpensive, fast, and readily provide answers to objective questions. Leased WATS and other special telephone services have reduced the cost of telephone surveys. The major cost is personnel.

The "yellow pages" are a source of lists for industrial wood products surveys. A company that manufactures components or special machines for wood cabinets can readily obtain telephone numbers from the yellow pages of each city or area. The *Thomas Register of American Manufacturers* lists telephone numbers of manufacturers by product categories. Directories of forest industry trade publications—for example, Miller Freeman's *Directory of the Forest Products Industry, Furniture Design & Manufacturing Annual Directory* issues, *The Red Book,* and directories of forest industry trade associations are sources of names for telephone interviews. State directories of industry list manufacturers by industries (sometimes by SIC codes.)

Mail interviews allow many people to be surveyed at one time at reasonable cost. As postage increases and the telephone becomes more competitive, the decision on dialing or mailing rests on the nature of the questions and the size of the survey sample. Mail surveys allow sufficiently large samples to deliver statistically significant results. Figure IX-4 is an example of how statistically significant survey results are obtained.

Personal interviews are the best method for obtaining detailed information, but they are slow and expensive. An experienced interviewer can gauge reactions by observing the body language of interviewees, which may reveal more information about feelings than verbal answers to questions.

Personal interviews are not utilized as frequently as they might be because householders are reluctant to open their doors to strangers. Market researchers are also reluctant to call on respondents at night, or to enter some neighborhoods.

Another type of survey is useful and affordable for smaller companies—*focus-group* interviews. Focus-group interviews bring together a small group of individuals who are carefully selected as potential customers of the product or service. Some focus interviews pay participants for their time. Others find that participants are so pleased to be included that paying for their time is irrelevant. One local newspaper conducted ten focus-group interviews among different types of readers. Even the busiest persons were happy to be asked to talk to the newspaper.

Ten people in a carefully run focus group can provide valuable insights for a business in an hour's discussion.

FIGURE IX-4

An Example of How Statistically Significant Survey Results are Obtained

Acquisition of statistically significant information is a science that uses defined terms and precise methodology. When professors Steven A. Sinclair and Robert J. Bush, of the Department of Wood Science and Forest Products, Virginia Polytechnic Institute and State University at Blacksburg, Virginia set out to determine the competitive strategies used in the U.S. Hardwood Lumber Industry, they followed this procedure:

1. They defined the problem with precision, focusing on discovering how business-level competitive substrategies address the problem of achieving and maintaining a competitive strategy among large U.S. hardwood lumber producers. Each part of the definition was, in turn, defined in scientific terms.

2. Based on a review of prior studies of competitive strategies used by wood-based industries, they then set up a model which identified strategic groups among large U.S. hardwood producers. Using 20 items which included variables such as developing new products, serving geographical markets, providing rapid delivery, they devised a questionnaire that rated the importance of each of the factors in determining company strategy.

3. The questionnaire was mailed to a sample of 80 of the 100 largest U.S. hardwood lumber producers (based on production), which were assumed to be influential firms.

4. An additional 19 companies were interviewed in person. This procedure resulted in a high 72 percent response to the questionnaire.

5. The answers were then analyzed using a three-factor analysis in which Factor 1 represented a differentiation strategy, Factor 2 was a focus strategy aimed at a specific market segment, and Factor 3 represented a cost leadership strategy. Cluster analysis—a technical term for classifying answers—was used to cluster companies into five separate groups.

6. The analyses concluded that the U.S. hardwood lumber industry was moving toward increased differentiation.

A focus group study for the Home Furnishings Council revealed that people now spend less time at home, but they value the time they do spend there more. Homes become their havens away from the pressures of work. The study concluded that for today's consumers emphasizing the pleasures of a home is a better focus than the status value of the home.

A manufacturer of safety equipment for automation control organized a focus group to obtain outside reactions to a new product the company was considering. Among the comments offered, the most helpful related to the price the company proposed to charge. The focus group thought it was too low. The company took their advice and put the new device on the market at a higher price than originally intended. The product sold well at the higher price.

Experiment

Experiments are conducted in market research when results with a test group can be compared with a control group. Test marketing—introducing a product to a test area and comparing the results with anticipated sales—is a form of experiment. Experimentation requires control of all variables, which is difficult when working with people.

Simulation

Simulation is a computer based technique for manipulating market factors. The validity of the simulation conclusions depends on the accuracy of the data inserted into a model and the assumptions on which the model is based. With the increased availability of consumer marketing information systems and PC programs, asking "what if" is becoming practical for market researchers.

Computer modeling based on econometric models forecasts consumer spending by studying prices, interest rates, income and wealth, demographics, and the confidence consumers have in the economy. By synthesizing data from millions of numbers, the econometric models attempt to get a handle on the larger economic factors that affect consumer behavior.

One population sector—the baby boom generation—is so large a group that it has had a major impact on spending for products. And the impact has varied as the group has aged. Today's boomers are growing more home oriented.

In a previous generation, this home centering may have predicted increased spending on furniture, or on coordinated ensembles and floor coverings. With more sophisticated tastes, more education, and two paycheck incomes, the maturing baby boomers are expected to choose a few pieces of quality furniture. The computer models also forecast that they will spend more on home entertainment equipment. These findings indicate that furniture manufacturers may do well to focus on entertainment centers and high-end furniture for this large market.

Companies specializing in programs for geodemographic information, consumer purchasing information, attitude and lifestyle information, and other marketing factors are almost daily adding to the roster of available information.

MARKET RESEARCH "HOW TO"

Market research is a useful tool for answering common business questions. The following sections show how the market research procedures discussed earlier can be applied to some typical business situations.

Gathering Information for Analyzing the Competition

Anyone planning to introduce a new product or service will meet head-to-head with competition from products and services already on the market. Information about the competition that the new product or service will face is one of the first tasks of researching a market. The strategies and plans of competitors are usually surrounded by a wall of secrecy, but a marketer does not need to be crafty to breach that wall. Information about competitors is available from the competitors themselves, from trade sources, and secondary sources—published information and government documents.

There are five simple, low cost methods for learning about the competition:

- Buy competitive products. If a company is considering entering the business of making decorative wooden bowls it would buy competitive wooden bowls available in the stores, compare their quality, packaging, and other features to determine the differential advantages of the planned product.
- Observe the competitors at work. Competitors themselves may provide more information about their company than they intend. Keen ears and sharp eyes help to pick up information from competitor's employees and facilities. Employees may inadvertently reveal that customer service has been exceptionally busy remedying problems in the use of a new product. The number of filled parking spaces may indicate if a plant is working to capacity. Noting delivery trucks may reveal the company's suppliers. Visits to competitors' facilities, listening to speeches by company executives, reading corporate publications (including annual reports, advertisements, press releases, and help wanted ads) reveals something about the status of competitors.

- ❑ Attend trade shows. New products and ideas are often first introduced at industry trade shows. They offer an opportunity to obtain a competitor's literature and a sense of the distribution and promotion strategy that the competition is adopting.
- ❑ Scan trade and professional publications. Publications can be mined for information on new competitive products, marketing strategies, and the latest industry rumors. The information gathered is as useful as the regularity and system with which it is obtained and recorded. Competitor information may be acquired informally, but is more valuable if the date and source of each piece of information is recorded. Noting advertising campaign themes for competitive products is a key to the trends that other companies have noted. Piggy back as often as feasible on the market research that has already been completed by competitive companies.
- ❑ Obtain secondary information from published sources. In addition to publications previously referenced in this chapter, other sources may be located from Gale's *Directory of Directories*, Gale Research Co., Book Tower, Detroit, MI 48226, or the *Business Periodicals Index*, published by H. H. Wilson Co., 950 University Ave., Bronx, N.Y. 10452.

Investigating the Market Potential for Expanding an Industrial Wood Product

A millwork company considers expanding sales by producing edge-glued moulding stock longer than other moulding manufacturers in the area. The company intends to provide just-in-time (JIT) delivery service. The company is already serving several steady customers. It wonders if the moulding stock market has room for expansion.

The latest *Final Industry Reports* of the Census of Manufacturers provides the data by Standard Industrial Code (SIC). SIC Code 24, lumber and wood products (except furniture), includes five general industries that might use moulding stock. The potential market for moulding stock is found in the SIC Code 243—Millwork, Veneer, Plywood and Structural Wood Members.

SIC Code 243 is subdivided into several groups:

- ❑ SIC 2431 Millwork
- ❑ SIC 2434 Wood kitchen cabinets
- ❑ SIC 2435 Hardwood veneer and plywood

- SIC 2436 Softwood veneer and plywood
- SIC 2439 Structural wood members, NEC (not otherwise classified)

SIC 2431, Millwork, is further divided into five digit SICs. SIC 24316 is wood mouldings. The company specializes in pine moulding stock. It would look under the seven digit classification SIC 2431621 for softwood moulding.

Under this classification the company finds that, in 1987, 93 companies made shipments of $100,000 or more. The census report also shows that the total value of softwood pine moulding shipped by these 93 companies was $799.9 million dollars. Consulting another table in the millwork section, the marketer calculates that moulding manufacturers consumed approximately $600 million worth of softwood lumber.

If a company intended to serve the western states market, it would consult the SIC 2431 data in Table 5 of the millwork census report to determine the value of moulding shipments from each of the Western states, which indicates the Western sector's share of the moulding market.

The 1987 data, when compared with the 1982 census data, shows that pine moulding sales more than doubled during that period. Any conclusion on growth would be tempered by remembering that 1982 was a recession year in the forest products industry, while 1987 was a good year.

Evaluating the Market Size for a New Product

The first port of call in estimating market size is the U.S. Census Bureau. Assume that a sawmill presently producing hardwood lumber is considering producing hardwood dimension specifically for the furniture industry in the geographic area where it is located.

Hardwood dimension would be purchased by manufacturers of wood furniture and upholstered wood furniture.

Looking up the SIC code 25—Furniture and Fixtures, the researcher would find three three-digit subclasses:

- Household Furniture (SIC 251)
- Office Furniture (SIC 252)
- Public Buildings and Related Furniture (SIC 253)

The data indicates that household furniture is subclassified into wood household furniture (SIC 2511) and upholstered hardwood furniture (SIC 2512).

Wood household furniture is further divided into five SIC subcategories:

- ❑ SIC 25112 Wood living room, library, family room and den furniture
- ❑ SIC 25113 Wood dining room and kitchen furniture, except cabinets
- ❑ SIC 25115 Wood bedroom furniture
- ❑ SIC 25116 Infant's and children's wood furniture
- ❑ SIC 25117 Wood outdoor furniture, unpainted wood furniture, and unassembled knock-down wood furniture

The hardwood lumber company can find even more specific data in the census. For example, assume that the prime interest is in SIC 26112—wood living room, family room, and den furniture (except sewing machine cabinets). The researcher would find these classifications for wood furnishings for living rooms:

- ❑ SIC 2511219 Cabinets
- ❑ SIC 2511233 Chairs (except dining room)
- ❑ SIC 2511235 Rockers
- ❑ SIC 2511241 Tables (except card and telephone tables)
- ❑ SIC 2511251 Desks
- ❑ SIC 2511271 Credenzas, bookcases, and bookshelves (except wall units)
- ❑ SIC 2511281 Wall units
- ❑ SIC 2511291 Other seating
- ❑ SIC 2511298 Other non-upholstered living room furniture

Assume further that a mill plans to supply hardwood dimension to manufacturers of SIC 26112 (wood living room, family room, and den furniture) on the West Coast. The census geographic area series is limited to the four-digit SIC 2511 (wood household furniture). Using the state of Oregon as an example, market research would discover that Oregon has 54 wood household furniture manufacturers. Only seven of these have 20 or more employees. Total shipments of Oregon wood household furniture manufacturers in 1987 were $48.0 million.

Another option for sizing up the market is input–output analysis, based on the realization that the output (sales) of one industry are the inputs (purchases) of other industries. The *Survey of Current Business,* published monthly by the U.S. Department of Commerce, includes the detailed input–output structure of the United States economy. The survey showed, for example, that industries in the SIC 2599 category (furniture and fixtures) sold almost half their total output in 1985 to eating and drinking places. Input–output analysis is time consuming and may best be left to market research experts skilled in using the technique.

Developing a Promotion Program

A small furniture company develops a line of sturdy, rustic furniture planned for second homes and weekend retreats and middle income families with young children. It intends to spend its limited advertising budget prudently. In what geographical area will advertising have the largest payback?

For assistance in this decision the company would consult *Sales & Marketing Management's* "Annual Survey of Buying Power—Part II." The survey provides detailed instructions on how to use the data. The furniture manufacturer might want to decide between spending advertising dollars in Tuscaloosa, Alabama or Medford, Oregon. The survey shows that Tuscaloosans spent almost 30 percent more on furniture than Medford residents. Advertising in Tuscaloosa should reach more ready customers than in Medford. Pretesting verifies the accuracy of this conclusion.

A MARKET INFORMATION LIBRARY

Unless a company has a marketing department that frequently requires detailed market information, acquisition of a market information library can become a costly office white elephant. There are some references, however, that are easy to use, relatively inexpensive, and can be readily acquired. Recommendations for a minimum market research library in the wood products industry are included in Figure IX-5.

DO-IT-YOURSELF OR PROFESSIONAL MARKET RESEARCH?

Somewhere in making marketing decisions, management comes face-to-face with the question of whether to undertake the market research in house or to retain a professional marketing consultant. The individual small business attempting to preserve scarce cash resources may have little choice but the do-it-yourself route in the initial stages.

Larger companies with in-house market research departments acquire their own secondary data and sometimes retain professional help to acquire and analyze primary data.

There are advantages in a hybrid approach. The business manager will benefit from spending some time talking directly with potential customers. Personally getting the feel of customer's needs and buying

FIGURE IX-5
A Minimum Market Research Library

- ❑ *U.S. Industrial Outlook* for the current year. Published annually by the U.S. Department of Commerce, Superintendent of Documents, Government Printing Office, Washington, DC 20402.
- ❑ Annual Buyer's Guides of any trade journals for industries in which the company has an interest.
- ❑ Current edition of *Statistical Abstract of the United States,* published by U.S. Department of Commerce, Superintendent of Documents, Government Printing Office, Washington, DC 20402.
- ❑ The monthly bulletin published by each Federal Reserve district. The 50 states are divided into 12 Federal Reserve districts. Check your local bank for the address of the nearest Federal Reserve Bank.
- ❑ *The Business Data and Market Information Source Book for the Forest Products Industry* compiled by Edwin Kallio and Edward Dickerhoff, published by the Forest Products Research Society, 2801 Marshall Court, Madison, Wisconsin 53705.
- ❑ *The Forest Industries Marketing Newsletter,* published four times a year by the Forest Products Research Society, 2801 Marshall Court, Madison, Wisconsin 53705.
- ❑ *Forest Industries Annual Directory,* published by Miller-Freeman Company, 600 Harrison Street, San Francisco, CA 94107.
- ❑ A current *World Almanac,* published annually by Newspaper Enterprise Association, Inc., New York, N.Y.

habits provides clues to the market demand, the best method of distribution, acceptable price ranges, and the type of promotion to use. These are clues, not answers. Answers require time and know-how best provided by professional marketers. Marketers also are able to analyze data with an objectivity the business manager usually lacks.

A company gains more value and more precise answers from market researchers if it knows clearly what's required and carries through recommendations. If the researcher recommends reorganization of the sales department and there's no way that department will be reorganized, money for research may not be well spent.

An experienced marketer provides answers to questions such as:

- ❑ Has the market been fully identified?
- ❑ Has the potential market been sufficiently narrowed for the company's size and range of skills?

- ❑ Is the product responsive to customers' needs, or only to the company's requirements?
- ❑ Have the differential advantages of the product been fully explored?

In addition, marketing specialists can perform a wide range of organizational audits that tell whether the company marketing structure and new product development procedures are adequate for its needs.

But no discussion of market research is complete without caveats. Data is too easy to misinterpret and misuse. Read the fine print in both public and privately published data. Note reporting changes. When new SIC classifications are added, old data included those products in another classification. Definitions are sometimes reworded. Put data in context. For example, comparing the 1982 census statistics with the 1987 data and 1977 data may give an entirely erroneous image of industry growth. The year 1977 was a boom year in the forest industry while 1982 was a recession year and 1987 was a very good year. Between 1977 and 1982 inflation soared to double digits.

Read the details of the research methodology. What companies were interviewed in a survey? If a market research survey interviewed only the 50 largest companies, are the conclusions relevant for smaller companies?

An example of misused research appeared in the February 1990 National Association of Exposition Managers (NAEM) newsletter.

The story read: "$373 million is the estimated result of an independent survey of the annual economic impact of the 47 year old Cleveland Home and Flower Show."

The story raised a furor among exposition professionals. The $373 million figure was clearly out of line and misleading.

The company that made the economic impact study had asked 625 of the attendees at the flower show which of several products they were reasonably sure of purchasing in the next 12 months. The products ranged from flowers to home improvements such as a room addition, decking, and siding. The survey also asked how much they anticipated their purchases would total.

The survey concluded that the actual direct economic impact of the attendees, exhibitors, and others on the City of Cleveland was about $6 million. How did this grow into $373 million? That figure was the anticipated economic impact. The survey came up with an average anticipated purchase figure, multiplied the average by the total number of attendees, and concluded that the Cleveland show had a $373 million potential. Nowhere did the report state that the purchases were, in fact, made.

The moral of the story is always study results with a skeptical eye! Note the methodology used in the research and conclusions.

Another source of misunderstanding are the terms used in market research. For example, there's a difference between household and family. A household consists of people who live in a housing unit. A housing unit is defined by the U.S. Census Bureau as a living unit with its own entrance. A family is a household that has at least two people who are related by blood, marriage, or adoption.

The term "traditional family" can be defined three ways: 1) a married couple family with their own children under age 18, 2) a family with a working husband and a non-working wife, or 3) a family with a working man, with a non-working wife and two children.

The first definition includes 26 percent of all households; the second brings the percentage down to 9 percent. The third comprises only 3 percent of American households. Nonfamily households are people who live in households but are not in families. These include single person households, unmarried couples, and roommates.

Race describes one of four racial groups: white, black, Asians and Pacific Islanders, and native Americans. Ethnicity refers to the nation or culture of origin. For example, Hispanic is an ethnic category referring to Spanish or Latin heritage.

Disposable income is after tax income. Discretionary income is the money remaining after people have paid for the necessities of life. Discretionary income indicates the market for luxury items.

X

THE ULTIMATE GOAL

. . . less risk, more profit

CHAPTER SUMMARY *A thriving business is the ultimate goal of marketing. Previous chapters have dealt with producing products and services that people want and need, as well as distributing, pricing, and promoting these products to win customers. The marketing program minimizes risks in introducing products and expanding markets.*

A well designed marketing program is an essential beginning for a thriving business, but converting the marketing idea to a profitable venture may fail when the costs of producing the product or providing the new service have been underestimated.

There are ten standard reasons why a venture may find that it has inadequate capital for a project. The four procedures frequently used to estimate capital requirements are compared and methods for determining the technological and economic feasibility of a new project explained. A realistic examination of anticipated investment and operating costs is necessary to minimize risks and increase the potential for profits.

Any discussion of new products and services or expanding markets ends with the bottom line question: "Can we make any money with it?"

The market studies discussed in this book minimize the risk of introducing the wrong products or services at the wrong time in the wrong way for the wrong reasons. Given a solid marketing investigation that encourages initiating the venture, owners and managers must now minimize the risks in investing in a new product. Marketable products are an essential beginning; profits with the venture are a function of the management skill and know-how of the business. Two companies may introduce a similar product at the same time and one may succeed while the other fails. The company that succeeds has management that knows its business, manages finances, people, and time efficiently, and can climb the growing mountain of government regulations. The successful company also makes satisfying its customers a highest priority. The business that lacks any of these skills jeopardizes profits.

But these management characteristics cannot assure success either if the business lacks adequate capital. If all the investment is spent before the plant starts up, management may never have the opportunity to use its skills.

Lack of adequate capital may be the result of the company's inability to obtain capital, or the company may have seriously underestimated the investment requirements and did not borrow adequate financing.

Shoestring ventures may succeed if the manager knows how to stretch the strings. The limitations imposed by not having sufficient money often encourage innovation and daring. The larger problem results when the company requests and receives financing of, say, $1 million, and it turns out that it needed $2 million.

Assuming that a comprehensive market study has been completed and that the project is judged feasible from a marketing perspective, a company arrives at the decision point where it must address the critical problem of determining how much capital will be required.

A company may be able to raise capital for the venture, but if it has not estimated the investment and operating costs accurately and conservatively it may end with a successful product and booming market but have to forfeit ownership in bankruptcy or forced sale because it lacks the capital to carry on.

A company may miscalculate the amount of capital it needs in any number of ways, such as:

- ❑ The projected operating costs are grossly underestimated
- ❑ The equipment selected is not based on sound technical analysis
- ❑ The plant layout is not engineered for maximum efficiency
- ❑ The technology for making the new product is not adequately developed
- ❑ The company is attempting to use new technology to produce saleable products without going through a pilot-plant stage
- ❑ Startup and debugging the production consumes the operating capital
- ❑ Technical production problems push the company into filling orders with products that fail to meet quality standards. Rejects and reruns reduce cash flow. Customers lose faith in company credibility
- ❑ Raw material costs are out of line with original cost estimates
- ❑ The venture is based on a preliminary or generic feasibility study which has missed important technical or economic factors
- ❑ Overlooked or new regulations add prohibitive costs

If the anticipated costs are so high as to diminish the projected profit so that the venture is no longer attractive to management, shaving anticipated costs is a temptation. Cutting back the scale of the operation, reviewing the manufacturing process, and engineering revaluation may reconcile the objectives of acceptable profit and lower investment.

Most of the above reasons for inadequate capital can be minimized by making a thorough feasibility analysis a determinant in the go/no go decision in a new venture.

THE FEASIBILITY ANALYSIS

Types of Feasibility Analyses

The word feasibility means practicability. A good feasibility analysis provides a risk assessment. Its purpose is to minimize predictable risks. An assessment may attempt to allow for unpredictable risks, such as inflation, economic downturns, long-term demographic changes, but such allowances are only as accurate as the crystal ball or tea leaves the analyst consults. Although these exogenous effects are not predictable, reasonably accurate critical technological and economic data is usually obtainable. This discussion is restricted to those factors that are predictable.

Feasibility analyses typically fall in four classifications:

- The quick and dirty type of analysis
- The generic project analysis
- The preliminary profit and loss (P/L) analysis
- The detailed site and situation specific analysis

Much of the difference between these types lies in the extent of assumptions. The quick and dirty type frequently is based on loose assumptions. The generic product analysis uses more, but general, hard data. The preliminary profit and loss analysis acquires as much hard data as possible without accounting for site and situation differences. The detailed site and situation specific analysis uses data valid for a specific operation at a specific site.

A feasibility study of a wood industry project would include a study of the technological feasibility and the economic feasibility.

A typical quick and dirty analysis assumes that the project is technically feasible and focuses on putting hoped-for numbers into an estimate of profit and loss. In contrast, a detailed study focuses on verifying assumptions of technological feasibility prior to analyzing economic feasibility.

The Quick and Dirty Analysis

The rough estimate of project investment and operating costs, often incorrectly called a feasibility study, and frequently made on the back of an envelope, has a rightful place in the early stages of planning as a "go or no go" indicator. Many successful businesses have been developed on the back of an envelope. Today's complex economy makes that type of quick estimate a reckless gamble. Although the figures may satisfy a business' desire for a tidy, quantitative evaluation, they are dangerous because the preliminary dollar figures frequently become the talking figure in discussions with financing sources.

The quick and dirty analysis may indicate that the project will require $1 million for new equipment and initial working capital. The quick estimate in hand, friendly financiers are persuaded to arrange a $1 million loan. Construction begins on the project; soon it becomes apparent that the quick estimate omitted some essential equipment and was much too low in other respects. The business soon becomes a another contribution to the statistics in undercapitalization.

The Generic Project Analysis

A seasoned executive is fully aware of the skimpiness of the figures on the back of the envelope. He remembers an article in a trade journal on a generic feasibility analysis for a product very similar to the one under consideration. The generic analysis appears to be quite thorough, with tables and footnoted data. A quick extrapolation and he acknowledges that the original $1 million dollar estimate should be $1.8 million. The $1.8 million becomes the talking figure with the financiers. Despite this extra care, the project could become another statistic in undercapitalization. What frequently happens is that the technological feasibility assumptions are considerably off the mark for the specific circumstances. The income is lower than projected and operating costs are higher.

The generic analysis may show that a hardwood dimension manufacturing plant may be feasible in a state, or that a particleboard plant may be profitable if located in the county conducting the analysis. To obtain a ball-park estimate, it is convenient to use available generic data based on general assumptions. When this generic analysis is the only basis for investment it spells danger, for it indicates only the potential for a venture to be profitable. It does not address the details unique to each business.

The Preliminary Profit and Loss (P/L) Analysis

An experienced manager goes a step further in determining the advisability of a project by making a preliminary profit and loss (P/L) analysis. The preliminary analysis is based on rough layouts, general estimates of equipment and productivity, income averages, estimates of operating costs, assumptions about sales, and personnel requirements. The preliminary analysis uses more hard data than most other analyses.

The additional research for the preliminary analysis may now show that the estimated cost should be $2.0 million. To reduce the cost of the preliminary analysis, investigation into the technological feasibility is usually brief. Frequently, that brevity is another cause of inadequate investment and operating capital.

Consider the analysis for planning a facility to manufacture mid-priced storage cabinets. The market study indicates that there should be a brisk demand for these cabinets. The product differential benefits have been designed into the cabinet. Distribution and promotion have been planned. But will the venture be profitable?

The facility will begin conservatively with room for expansion. According to the market study, the market would buy about 80 cabinets a day at an F.O.B. factory price of $95 each. A list of machinery requirements is drawn and the person designated as project manager phones a potential supplier for a cost estimate. The equipment supplier provides a talking figure of about $70,000. The $70,000 is entered as the investment. The project manager locates a suitable building available for lease. Construction expense should be modest. The equipment supplier reckons that the equipment estimated should be able to produce the 80 cabinets with about 10 people on the floor.

Factoring in projected operating costs, personnel expenses, raw material and sales costs yielded a bottom line of $215,000 in earnings before taxes for the first year of operation. On that basis, the project is given the go ahead.

But the $70,000 investment in machinery and equipment could produce at best, only 50 complete cabinets a day and requires the addition of two more operators to achieve that level of production.

The meticulous cash flow projections, the neat calculations of net present value (NPV) and internal rate of return (IRR) were of little purpose! The basic premises did not hold. The cabinet manufacturing facility proved to be an example of skimping on technological feasibility and leading the investors down a costly path.

Experience has shown that profitability estimates most frequently suffer from inadequate information on material costs and availability. Log sizes may be higher or lower than assumed. Available grades might be lower or degrade may be higher.

Construction overruns, often due to inadequate analysis of equipment requirements, resulting in low estimates of equipment investment costs also hurt profitability estimates. Operating cost overruns and overly optimistic assumptions of production capacity of equipment selected, resulting in unattainable production estimates can also contribute. Other factors may be too few employees to meet production requirements or the costs of "bugging out" and start-up have not been realistic.

Experience suggests that pages of economic analyses make convincing presentations, but they mean little if a sound technological analysis has not been made first.

The Site- and Situation-Specific Analysis

The most reliable feasibility analysis is "bankable" because it is the most credible. The bankable analysis minimizes risk by using hard data, painstakingly gathered, in place of assumptions wherever possible. It is specific for the site selected.

For example, a construction site on a steep hill or without road access may double the cost of site preparation. Transportation in the area selected may be poorly served, increasing transportation costs. Or projected labor rates may run higher due to the unavailability of affordable housing in the area. Competition for timber resources may push raw material costs higher than estimated.

Standard logging practices in the area may not yield the projected log lengths, requiring an additional operation at the processing plant.

Rigorous checking of technical and economic data marks the site-specific feasibility analyses. The bankable analysis does not eliminate risk from unpredictable events, but it minimizes risk from factors that can be known. That knowledge is supported by conducting a technological analysis and an economic analysis.

The technical analysis reveals problems that may arise due to technological factors that, if not addressed, could spell disaster in a manufacturing venture. Some of the questions that might be raised are listed in Figure X-1.

The questions in Figure X-1 represent some of the technical and management considerations that are almost routinely responsible for cost overruns, and profit underruns. It may also reveal opportunities to give the new facility a competitive edge.

What will the "bankable" feasibility study reveal? The credibility of the site-specific analysis rests on its objectivity and rigorous examination

FIGURE X-1
A Worksheet on Risk Factors in a New Venture

Enter "yes" or "no" in the spaces next to each question.

_____ 1. Is the company familiar with marketing the type of product?

_____ 2. Is the company experienced in manufacturing the type of product?

_____ 3. Will the manufacturing process require application of a new technology or new equipment?

_____ 4. Have other processors used this technology or equipment?

_____ 5. Are equipment capacities used in the feasibility study based on experience?

_____ 6. Has a pilot plant been constructed?

_____ 7. Are the equipment costs and capacities based on sound engineering design and calculations?

_____ 8. Are the equipment costs and capacities based on machinery suppliers' quotations?

_____ 9. Is the process flow plan standard?

_____ 10. Is there a provision for "bug out" costs in using new equipment?

_____ 11. Will anticipated revenues pay for the development costs?

_____ 12. Are provisions being made to minimize start-up cost-overruns?

_____ 13. Is the wood resource available and accessible?

_____ 14. Is there likely to be a problem in the near future with public pressure to use the resource for other purposes?

_____ 15. Is the desirable species available in the quality and quantity required?

_____ 16. Can alternative species be utilized?

_____ 17. Are trained employees available?

If there are many "no" answers, use this worksheet as a reminder to convert them to "yes" answers. Or, allow costs for remedying the "no's" in the economic analysis.

of assumptions. The analysis should provide financial sources with an understanding of:

- ❑ The technical and economic factors considered
- ❑ The status of the equipment to be used. Is the equipment developmental, experimental, or standard?

- ❑ The anticipated time and costs for start up
- ❑ Anticipated "bug out" problems
- ❑ Questions that have been asked and answered
- ❑ Additional research and development requirements and funding
- ❑ The data sources
- ❑ Conservative estimates of potential profits
- ❑ Availability and cost of raw material (a critical aspect in a natural resource-based industry)
- ❑ Sensitivity of the process to energy costs, transportation, labor requirements, and similar elements of operating costs

Assumptions about the availability and cost of the raw material resource are often troublesome. If the raw material has not been secured by contract, the estimated material cost may be out of line. How reliable are the estimates of the resource? Has the timber been cruised? A manufacturing facility that expands on the assumption that additional resource is available may be forced to operate only one shift when profit projections were based on two.

After the technical questions have been resolved and dollar signs attached to technical considerations, the stage is set for the economic feasibility study. With anticipated costs, can this product make a profit for the manufacturer?

The bottom line of economic feasibility is the profit and loss (P/L) statement. Specific data for those elements of income and fixed and variable expenses that determine profit or loss may produce a picture entirely different from rough estimates. To produce that hard data someone has to first go to the drawing board and calculator. The data is only as accurate as the effort and experience that has gone into the planning. An economic analysis should consider raw material input, anticipated product output, anticipated investment requirements, labor costs, and plant operating costs.

The anticipated product output is the production in appropriate units (i.e., lumber in board feet or number of units.) Output includes not only the desired premium grades or units of items to be produced, but also material that can't be utilized in the major products. Consider trim ends, bark, sawdust, and other by-products. If they are to be handled as waste, factor in a waste disposal cost. If they are to feed a profit center such as chip sales, or be a cost reduction center, such as generating steam for dry kilns or electricity, estimate income from this source.

Success in manufacturing secondary products (e.g., pallet stock and furniture components) depends on deriving income from all the material purchased. The sales values of products from what might be considered residue in a primary mill may make the difference between profit and loss in manufacturing secondary products.

Calculating anticipated sales prices tests the usefulness of any Ouija board or crystal ball. It is the principal variable in estimating income. Several methods are commonly used for estimating sales prices: rolling averages of sales prices for the past five years, extrapolating future income from a graph of historical prices, calculating the impact of a historical inflation average, and averaging a series of high and low prices.

The anticipated investment requirements are the basis for depreciation costs, the investment hinges on the design concept of the manufacturing facility. For example, one manufacturer's engineered production flow plan shows that processing will require a dry kiln, a planing mill, remanufacturing facilities, a dust handling system, finishing equipment, and a boring machine. Equipment manufacturers are helpful in providing quotations, but their price quotations are only as accurate as the specifications and performance capacities provided to them. Equipment performance capacities, power and installation requirements are calculated with care. Any one machine that cannot keep up with the performance of the other machines slows production. Can the product be readily transferred from one machine to another in processing? What are the dimensions, weights, power requirements, and foundation requirements?

With the dimensions of equipment in hand, building and site requirements (i.e., acreage, terrain, utilities) are calculated. Contractors use building and site dimensions to quote construction and site preparation, including required permits and real estate costs. Some sites require bringing in water, constructing roads, and accessing other utilities.

Labor costs are a critical variable in the cost of operation. How many people will be required to run the equipment for the anticipated production? What levels of skill are required?

A production flow plan might indicate the number and type of personnel required at each position. Personnel costs may represent the largest element of operating costs. Escalating fringe costs for benefits unheard of a few years ago, such as health insurance, child care, parental leave, and soaring workman's compensation, turn every additional person into a significant cost. Production rates quoted by equipment manufacturers should be checked with several users of the equipment, if possible.

Operating costs include labor, fringe benefits, taxes, utilities, insurance, waste and pollution control, and similar operating expenses. Operating costs are, in turn, divided into fixed costs, which continue even when the plant is not operating, and variable costs, which vary with the scale of plant operation. Figure X-2 describes fixed and variable operating costs.

THE PROJECTED ANNUAL PROFIT/LOSS STATEMENT

Figure X-3 is a form for a typical projected profit and loss statement. Income less total costs equals apparent profit or loss before taxes. After management is satisfied that the profit projection is based upon the most accurate data available, cash flow is considered. Net present value (NPV) and the internal rate of return (IRR) are accounting considerations that answer for each manager the potential income that can be earned in a new venture compared with potential earnings from investing in the stock market or bonds, or a another venture.

The discounted cash flow technique recognizes that money has a time value. A dollar in hand today is worth more than a dollar received tomorrow. The net present value (NPV) is the value of future cash inflow compared with the present value of the cash outflow at a specified interest rate. Whenever the net present value is greater than zero, the project is considered economically feasible.

Profit and loss statements can be set up on a computer program to compare options and answer "what if" questions: What if prices drop? What if the cost of raw material increases? These "what if" questions allow managers to examine the sensitivity of proposed ventures to influences beyond their control.

FIVE MISTAKES TO AVOID

Financial judgments are often befogged by knee jerk reactions and mental blind spots. Investments in new ventures suffer from the same pitfalls. Five mistakes to avoid have been identified by financial experts:

- ❑ Overconfidence. Data is assumed to be correct, even when the information is sketchy. Separate the known from the unknown.
- ❑ Using inadequate data to make important investment decisions.
  Double check estimates of the most significant factors.
  Underestimating personnel or overestimating sales quickly turns a profit into a loss.

FIGURE X-2

Fixed and Variable Operating Costs in a Wood Products Manufacturing Plant

Variable costs rise and fall with the production level and operating hours. They include:

❑ Maintenance costs. These are of increased significance as equipment becomes more technologically sophisticated.

❑ Operating supplies such as saw blades and lubricating oil

❑ Consumables such as nails, staples, and paint

❑ Utilities, including water, gas and electric power, and sewer costs

❑ Miscellaneous costs, such as spraying logs in summer, protecting logs from stain, and humidity control for dried lumber

❑ Overhead, which includes employers contributions to health, welfare, workman's compensation, paid vacation, and paid holidays

❑ Waste disposal costs such as landfill, dust control, and pollution control

❑ Liability insurance scaled to employee hours and product insurance based on production levels

❑ Sales commissions, discounts, and trade association dues that are tied to sales volumes

Fixed costs continue when the plant is not operating. These include:

❑ Depreciation on machinery and equipment

❑ Fire and basic liability insurance

❑ Property taxes

❑ Repayments and interest on borrowed money

❑ Other miscellaneous costs for things such as utility hook-up, fire patrol, and security

❑ Permits, licenses, association dues

❑ Contingencies

Sales and management costs, are partially fixed and partially variable.

Sales and management or administration costs for paid salesmen, plant managers, and other management and office personnel are fixed costs. These costs and sales promotion costs are somewhat flexible. When business is slow some office personnel and salesmen may be laid off. Advertising may be curtailed.

FIGURE X-3

A Worksheet For Calculating A Projected Profit and Loss (P/L) Statement

PROJECTED ANNUAL PROFIT/LOSS PRO FORMA STATEMENT

A. ESTIMATED INVESTMENT

B. INTEREST RATE %

C. LOAN PERIOD YEARS

D. SALES INCOME

1. PRODUCT 1
2. PRODUCT 2
3. PRODUCT 3
4. RESIDUE
5. TOTAL SALES INCOME

E. OPERATING COST

1. LABOR
2. MAINTENANCE AND OPERATING SUPPLIES
3. UTILITIES
4. GENERAL OVERHEAD
5. INSURANCE
6. OPERATING CONTINGENCY
7. TOTAL OPERATING COST

F. RAW MATERIAL COST

G. TOTAL PRODUCTION COST

H. SELLING AND MANAGEMENT COST

I. TOTAL COST

J. NET INCOME

K. LESS DEPRECIATION

L. LESS INTEREST COST

M. NET BEFORE TAX PROFIT

APPARENT RETURN ON INVESTMENT

RETURN ON INVESTMENT

RETURN ON SALES

- ❑ Paying too much attention to memorable events that may cloud judgment. The local media may headline the failure of a window manufacturing company located in a small town. That failure doesn't mean that all window manufacturing plants are unusually risky ventures. Determine why the company failed before considering the failure as a market warning.
- ❑ Ignoring the odds. Try to understand, rather than dismiss, discouraging events.
- ❑ Hanging on to an investment too long if there is a potential for further loss. Conversely, don't let early losses due to start up problems precipitate a premature decision that the investment was a mistake.

Successful businesses have been started on hunches and backs of envelopes. Investing in planning and engineering, and time consuming data gathering may appear too cautious. But in assessing the value of planning compared with instincts, businesses often ignore expensive remedies such as the legal fees for litigation against equipment manufacturers who provided inadequate equipment, or the long delays in starting up when defective flow plans are barriers to reaching production objectives. In today's business environment, rapidly advancing technology, skyrocketing labor costs, and short supplies of skilled labor make investments in manufacturing plants more risky than in prior eras. Business needs all the help it can get to minimize risks. Shooting from the hip may have worked yesterday but no business can afford them today.

APPENDIX A

. . . a glossary of marketing terms

Accessory Equipment—Less expensive and shorter lived assets, such as office machinery and hand tools, used in industry.

Advertising—Presentations in the media to call public attention to a product or service with the objective of stimulating sale. The advertiser is usually identified in the advertisement and usually pays for the space or time used. See also comparative advertising, institutional advertising and speciality advertising.

AOI Statements—Statements used in psychographic studies to connote activities, interests, and opinions of study participants.

Brand—An identifying name, term, symbol, design, or combination of these that the consumer associates with only a single product, product line or selling company.

Broker—An independent wholesaler in the distribution system who does not take title or possession of the products he brokers, but works primarily to bring buyers and sellers together.

Cash and Carry Wholesaler—One who performs most wholesaling functions except financing and delivery. See also wholesaler.

Comparative advertising—Selling efforts that promote products by making direct comparisons with other competitive brands. See also advertising.

Catalog Retailer—a merchant who distributes from a showroom in which samples of a product line are displayed. Customer's order from the catalog and orders are filled from a warehouse.

Celebrity Testimonials—Statements from well-known entertainment, sports, political, or other hero-figures testifying to the value of products.

Channels—A marketing term designating the paths that goods, and title to them, follow from the manufacturer to the consumer.

Channel Captain—The member of a marketing channel who is dominant and controlling.

Channels of Distribution—A method employed by a company to distribute its products through wholesalers, distributors, jobbers, retailers or other middlemen.

Commodity—An item of merchandise not distinguishable from like items either as to origin or use.

Community Shopping Center—A group of 15 to 50 retail stores, often including a branch of a department store as the primary tenant. This type of center typically serves 20,000 to 100,000 persons within a radius of a few miles.

Competitive Position—A company's place relative to its competitors, may be expressed as market share, or percentage of total product sales.

Concentrated Marketing—Focusing of all of a firm's marketing resources into a small segment of the market.

Consumer—A customer who buys a product or service, usually an end product, for his own personal use. Commonly used in reference to individuals rather than industrial, institutional, or other classes of consumers.

Consumer products—Products produced for direct use by individuals and households by purchased the ultimate customer for personal use.

Convenience products—Standard products that consumers purchase frequently, with minimum effort, and from the most convenient source.

Cost Plus Pricing—The practice of deriving the selling price of an item by marking up a base cost figure with overhead costs and a profit.

Coupon—A sales promotion method using a specially marked coupon that entitles the bearer to a discount on the purchase of a particular product.

Customer—An individual or organization that purchases a product. May be a consumer of the end product for personal use or industrial buyer purchasing for intermediate use in making products.

Demand Variability—In the industrial market, the impact of derived demand on the demand for component products or materials used in producing end products.

Demographics—Age, sex, income level, and behavior characteristics of potential buyer groups. The study of human population densities, distributions and movements.

Demographic Segmentation—Dividing a population into groups that share the same age brackets, sex, income level, or othe characteristics. See also Segmentation.

Demonstrations—A display or explanation showing how a product or concept works or is used.

Derived Demand—In the industrial market, the demand for an industrial product that is linked to demand for a consumer item.

Differential Advantages—Benefits of a product over and above those of competitive products.

Differentiated Marketing—A technique employed by a firm that produces numerous products with different marketing mixes designed to satisfy smaller market segments.

Differentiation—Product attributes or benefits or services that differentiate a product from its competitors.

Differentiation Benefits—A package of benefits or attributes that increases the value of a product with respect to competitive products.

Differentiation Strategy—A marketing strategy based on the differential advantages of a product.

Distribution—The system by which a company gets its products to the point of purchase by the final consumer.

Distribution Marketing Channels—The steps a product or service follows from producer to final consumer.

Driving Force—The primary determining factor behind marketing decisions regarding major products and markets. The basis of strategic management decisions.

Drop Shipper—One who receives orders from customers and forwards them to producers who ship directly to the customer.

Experiential Pespective—The beliefs, feelings, or emotions that influence buying decisions; a significant factor in positioning products.

Fabricated Parts And Materials—Finished industrial products that become part of the final product (i.e. dimension components for the furniture industry.)

Focus Group—A marketing research information gathering procedure that typically brings several individuals together in one location to discuss a given subject.

Franchising—A system of contractual arrangements by which a dealer complies with the operating requirments of a manufacturer.

Forward Integration—Manufacturers who also act as distributors of their products.

Geographic Segmentation—Dividing a population into buying groups on the basis of location. See also Segmentation.

Impulse Buying—Consumer purchases which take little, if any, deliberation.

Impulse Products—Products purchased by a customer on the spur of the moment, with little or no conscious deliberation.

Industrial Products—Products used either directly or indirectly in production of other products.

Industrial Customers—Customers who purchase products and services for use in producing other products for resale such as manufacturers, retailers, wholesalers, contractors, or construction firms.

Institutional advertising—Promotion of an idea, concept, philosophy, or goodwill, of an industry, company, or organization. See also advertising.

JIT—See Just In Time.

Joint Demand—In the industrial market, the demand for an industrial product or service as related to the demand for another industrial product that is necessary for the use of the first item.

Just In Time—A manufacturing practice and marketing concept that assures the arrival of every part at the right time and right place just in time for assembly.

Lifecycle—The stages of a product sales life. There are four generally accepted life cycle phases: (1) development, (2) growth, (3) maturity, and (4) decline. The duration and intensity of each phase depends upon the characteristics of each product or business.

Lifestyles—The mode of living of consumers.

Manufacturer's Representative—An independent salesperson who represents a manufacturer and receives a commission based on products sold. The representative does not take possession of the goods, does not provide credit, and does not inventory or deliver products.

Merchandising—The act of buying and selling, carrying on trade in products.

Market—A segment of the population with purchasing power and the ability and willingness to buy, such as the youth market, mature market, or Hispanic market. Also refers to a segment of industry, such as manufactured homes or furniture.

Marketing—The development and distribution of products and services for selected population or industrial segments.

Marketing Channels—The steps that products or services follow from producer to reaching the final consumer.

Marketing Mix—Coordinating products, pricing, distribution, and promotion, to satisfy the needs and wants of selected customer segments.

Market Research—Systematic gathering, recording, and analysis of information about the marketing of products and services.

Market Segment—A portion of a total market with similar product interests or other characteristics.

Market Segmentation—The process of dividing the total market into smaller, more homogeneous groups with similar product interests, based on any number of characteristics. See also Segmentation.

Market Targets—The segment of the market selected by the manufacturer as the focus of marketing activities.

Middleman—A business firm that operates between the producer of products and services and the consumer or industrial user.

Mission—The special task or function on which a business focuses its efforts.

MRO Items—Supplies for an industrial firm categorized as maintenance items, repair items, or operating supplies.

Need—Something useful or required, a necessity, an urgent requirement.

Neighborhood Shopping Center—A cluster of stores, usually consisting of a supermarket and smaller stores, that provide convenient shopping in the geographical vicinity.

Niche Marketer—A manufacturer that serves a narrow segment of the market better or more efficiently than any other company.

Niche Product—A product that meets the needs and wants of a specific market segment.

Overall Cost Leadership—A strategy in which a firm aims to capture a large share of the market by virtue of lower prices.

Ownership Utility—A want-satisfying power created by marketers when title to a product is transferred to the customer at the time of purchase. See also Utility.

Personal Selling—Selling on a person to person basis. Any form of oral sales presentation or assistance to a customer.

Place—Satisfying customer wants by having products available where customers want to buy them.

Place Utility—A want-satsifying power created by marketers who have products available where customers want to buy them.

Planned Shopping Center—A unit for shoppers in a geographic trade area consisting of a planned, coordinated, and marketed group of retail stores.

Point Of Purchase (POP)—Advertising, displays, and other promotions at locations where the customer makes the decision to purchase.

Positioning—A marketing strategy to promote differential product advantages aimed at a specific segment of the market and designed to position the product in the mind of a prospective buyer.

Premium—Items given free with the purchase of a product or service.

Premium Products—Products that satisfy a want for prestige and high quality, usually higher priced.

Price—The amount the buyer is willing to pay for a product. The value a seller sets on the product.

Price Elasticity—An economic concept that attempts to measure the sensitivity of demand for any product to changes in its price. If consumers are relatively sensitive to changes in price (i.e., higher price results in a lower demand for a product), then the demand for such a product is elastic. If changing price does not affect the quantity purchased, then such a product exhibits an inelastic demand characteristic.

Price Lining—The practice of identifying market segments to which a company is appealing and setting a limited number of prices.

Product—An identifiable item, a package of benefits related to physical, service, or other attributes and designed for a specific use for a specific market segment. In popular terminology product refers also to services, securities and the like.

Product Differential—Presenting a product so that it is perceived by customers as unique or somewhat apart from other products available.

Product Lifecycle—The stages in the life of a product. See also Lifecycle.

Product Portfolio Analysis—A method for examining pricing and other management decisions based on the assumption that cash flow and sales volumes are closely related to sales volume. Also known as the Boston Consulting Group model.

Promotion—Actions taken to inform, persuade, and influence the customer's decision to purchase. Promotion can refer to various forms of advertising, trade or consumer discounts, specially advertised discounts, and similar promotion techniques.

Promotional Strategy—A blend of personal selling, advertising, publicity, and sales promotion to achieve sales objectives.

Promotionally Priced—A lower than normal price used as part of a firm's selling strategy.

Psychographics—Development of profiles of behavior from analyses of activities, opinions, interests, and lifestyles used to segment consumer markets.

Psychographic Segmentation—Behavior profiles developed from analyses of activities, opinions, interests, and lifestyles used to segment consumer markets.

Psychological Positioning—Use of advertising, public relations, point of purchase techniques, and other communication methods to move a potential purchaser toward a buying decision.

Publicity—Promoting products or services without identifying the sponsoring company. Indirect stimulation of demand for a product or service by arranging for significant news about it to appear in a published medium or obtaining favorable presentation of it on radio, television, or stage that is not paid for by the sponsor.

Raw Material—Farm products and natural resources used in producing the final manufactured products.

Regional Shopping Center—The largest type of planned cluster of retail stores, usually involving one or more major department stores and many other stores, and typically located in an area with at least 250,000 people.

Reverse Channels—Moving used products or by-products from the consumer to a manufacturer for recycling.

Sales Promotion—Displays, shows, expositions, demostrations, and marketing activities (other than personal selling, advertising, and publicity) to stimulate customers purchasing and dealer activities. In retailing, all methods of stimulating customers purchasing, including personal selling, advertising, and publicity.

Scrambled Merchandising—The practice of a retail store carrying dissimilar lines of merchandise in an attempt to produce additional sales volume.

Secondary Products—Products made from lumber or panelboard or other partially processed wood material in a form that can be used without futher processing by consumers or industries, as distinguished from primary commodities such as lumber, which results when logs or fiber raw materials are processed.

Segmentation—Dividing the population into potential buyer groups. See also Demographic Segmentation, Geographic Segmentation, Market Segmentation, and Psychographic Segmentation.

Selling—The exchange of products, services, or ideas in which a buyer acts favorably upon something that has commercial significance to the seller.

Selling Agent—A middleman who assists or persuades a prospective customer to buy a product or a service.

Sensitivity—The influence on sales and profits of conditions over which a business has no control.

Shopping items—Products purchased after a customer has made comparisons of features such as price, quality, and style between competitive items. Shopping goods such as furniture and rugs are often purchased infrequently, consumed or used up slowly, and are subject to deferred purchase decisions on the part of the customer.

Singularization—Adding value to wood products by differentation on the basis of the exceptional qualities, uniqueness, interesting origin, or other characteristics that distinguish one wood from another.

Skimming—Using a pricing policy which sets a high entry-price, allowing the manufacturer to quickly recover new product costs.

Specialty advertising—A sales promotion method that uses articles such as key rings, calendars, and ballpoint pens bearing the advertiser's name, address, and advertising message to reach the consumer target. See also advertising.

Specialty Products—Products purchased for unique characteristics that fill special wants of the customer, usually higher priced and branded. Also called premium products.

Standard Industrial Classification (SIC)—A classification system used by the U.S. Government to collect and disseminate information about industries.

Strategic Planning—The process of determining an organization's primary objectives, allocating funds and planning a course of action to achieve those objectives.

Strategy—A plan representing the intent of a company to achieve a particular objective or objectives.

Supplies—Regular expense items, such as office stationery, necessary for daily operations, but not part of the final product.

Target Market—A well defined group of potential customers that a company has selected to serve.

Tertiary industries—Business that specialize in producing services.

Truck Wholesaler (Truck Jobber)—One who markets perishable food items. Also known as a truck jobber. See also Wholesaler.

Utility—An economists term for the want satisfying power of a product or service. See also Ownership Utility and Place Utility.

Value Added—The difference between the price charged for a product and the cost of raw materials, labor, and other costs. Also, value added by marketing—the part of the value of a product or a service to the consumer or user which results from marketing activities.

VALS System—A values and lifestyles system devised by Stanford Research Institute International (SRI) to classify consumers into behavior types.

Want—A term used in marketing to describe some product or service or value that people wish for or desire, as contrasted with Need which is a requirement or necessity.

Wholesaler—A distributor who makes quantity purchases from manufacturers (or other wholesalers), and resells in smaller quantities, and takes title to the products handled. Also called wholesaling middleman. See also Cash and Carry Wholesaler and Truck Wholesaler.

APPENDIX B

. . . directory of forest industry organizations

Including Organizations Related to the Forest Industry

American Forest Institute
1619 Massachusetts Ave. NW, Washington, D.C. 20003

American Furniture Manufacturers Assn.
918 16th St. NW, Ste. 402, Washington, D.C. 20006

American Furniture Manufacturers Assn.
Box HP-7, High Point, NC 27261

American Hardboard Assn.
520 N. Hicks Rd., Palatine, IL 60067

American Institute of Timber Construction
11818 S.E. Mill Plain Blvd., Ste. 415. Vancouver, WA 98684

American Ladder Institute
111 E. Wacker Drive, Suite 600, Chicago, IL 60601

American Laminators Assn.
419 Norton Building, Seattle, WA 98104-1584

American Lumber Standards Committee
Box 210, Germantown, MD 20874

American Pulpwood Association
1025 Vermont Avenue, NW, Suite 1020, Washington, D.C. 20005

American Wood Council
1619 Massachusetts Ave. NW, Suite 500, Washington, D.C. 20036

American Wood Preservers' Institute
1945 Old Gallows Rd., Ste. 550, Vienna, VA 22182

Architectural Woodwork Institute
2310 South Walter Reed Drive, Arlington, VA 2206-1199

Association of Wall and Ceiling Industries International
1600 Cameron St., Alexandria, VA 22314

Building Research Board
2101 Constitution Avenue, NW, Washington, D.C. 20418

Ceilings & Interior Systems Contractors Assn.
5700 Old Orchard Rd., Skokie 60077

Communications Channels, Inc. (Fence Industry)
5 Penn Plaza, New York, NY 10001-1810

Decorative Laminate Products Assn.
600 S. Federal St., Suite 400, Chicago, IL 60605

Doors and Hardware Institute
14170 Newbrook Dr., Chantilly, VA 22021-2223

Fine Hardwoods-American Walnut Assn.
5603 W. Raymond, Suite O, Indianapolis, IN 46241

Forest Industries Council on Timber Valuation & Taxation
1250 Connecticut Avenue NW, Suite 230, Washington, D.C. 20036

Furniture Manufacturers Assn.
1500 E Beltline SE, Grand Rapids, MI 49506

Hardwood Institute
Division of National Hardwood Lumber Assn.
Box 34518, Memphis, TN 38134

Hardwood Manufacturers Assn.
400 Penn Center Blvd., Suite 530, Pittsburgh, PA 15235
(412) 829-0770

Hardwood Plywood Manufacturers Assn.
1825 Michael Faraday Drive, Box 2789, Reston, VA 22090

Hickory Handle Association, Inc.
c/o Donald Cowan
2923 22nd Avenue S, Nashville, TN 37215

International Hardwood Products Assn. Inc
Box 1308, Alexandria, VA 22313

International Sleep Products Assn.
333 Commerce St., Alexandria, VA 22314

Laminating Materials Assn., Inc.
767 Park Ave., Oradell, NJ 07649

Maple Flooring Manufacturers Assn.
60 Revere Dr., Suite 500 Northbrook, Chicago, IL 60062

Material Handling Institute Inc.
1326 Freeport Road, Pittsburgh, PA 15238

National Association of Home Builders
15th and M Streets NW, Washington, D.C. 20005

National Association of Housing and Development Officials
Watergate Building, Suite 404, 2600 Virginia Avenue, NW
Washington, D.C. 20037

National Association of the Remodeling Industry
1901 N. Moore St., Ste. 808. Arlington, VA 22209

National Dimension Manufacturers Assn.
1000 Johnson Ferry, Suite A 130, Marietta, GA 30068

National Forest Products Assn.
1250 Connecticut Ave. NW, Washington, D.C. 20036

National Hardwood Lumber Assn.
Box 34518, Memphis, TN 38184

National Kitchen Cabinet Assn.
P.O. Box 6830, Falls Church, VA 22046

National Lumber & Building Material Dealers Assn.
40 Ivy Street SE, Washington, D.C. 20003

National Oak Flooring Manufacturers Assn.
P.O. Box 3009, Memphis, TN 38173-0009

National Paint & Coatings Assn.
1500 Rhode Island Avenue NW, Washington, D.C. 20005

National Particleboard Assn.
18928 Premiere Ct., Gaithersburg, MD 20879

National Retail Hardware Association
5822 W. 74th Street, Indianapolis, IN 46278

National Roofing Contractors Association
10255 W. Higgins, Suit 600, Rosemont, IL 60018

National Unfinished Furniture Institute
1850 Oak St., Northfield, IL 60093

National Wood Window & Door Assn.
205 W. Touhy Avenue, Des Plaines, IL 60018

National Wooden Pallet & Container Assn.
1800 N. Kent, Suite 911, Arlington, VA 22209-2109

Railway Tie Assn.
314 N. Broadway, St. Louis, MO 63102

Society of American Wood Preservers Inc.
7297 Lee Higway, Unit P, Falls Church, VA 22042

Southern Pine Inspection Bureau
4709 Scenic Highway, Pensacola, FL 32504

Structural Insulated Panel Association
c/o Steven Winter & Associates
1090 Vermont Ave. NW, Suite 1200, Washington, D.C. 20005

The Associated Cooperage Industries of America Inc.
2100 Gardiner Lane, Suite 100E, Louisville, KY 40205-2947

The Business and Institutional Furniture Manufacturers Assn.
2335 Burton St. SE, Grand Rapids, MI 49506

The Formaldehyde Institute
1075 Central Park Avenue, Scarsdale, NY 10583

The National Decorating Products Association of America
1050 North Lindbergh Boulevard, St. Louis, MO 63132

The National Lumber Exporters Assn.
805 Sterick Building, Memphis, TN 38103

Walnut Council
5603 W. Raymond St., Suite O, Indianapolis, IN 46241-4390

Wood Products Manufacturers Assn.
52 Racette Ave., Gardner, MA 01440

APPENDIX C

. . . list of forest industry trade journals

Automated Builder
4371 Carpinteria Ave.
Carpinteria, CA 93013

CabinetMaker
Delta Communications Inc.
400 N. Michigan Avenue, Chicago, Illinois 60611
(312) 222-2000

Construction Dimension Magazine
Association of Wall and Ceiling Industries International
35 K Street, NE, Suite 300, Washington, D.C. 20002
(202) 783-2924

Decorating Retailer
The National Decorating Products Association of America
1050 North Lindbergh Boulevard, St. Louis, MO 63132
(314) 991-3470

Doors and Hardware
Doors and Hardware Institute
7711 Old Springhouse Road, McLean, VA 22102
(703) 556-3990

Dry Retailing
National Retail Hardware Association
770 North High School Road, Indianapolis, IN 46224
(317) 248-1261

Export Today
Trade Communications, Inc.
733 15th St., NW, Suite 1100
Washington, DC 20005

Fence Industry
Communications Channels, Inc.
390 Fifth Avenue, New York, NY 10018
(212) 613-9700

Forest Industries
Miller Freeman Publications
600 Harrison Street, San Francisco, California 94107
(415) 905-2200

Forest Products Journal
Forest Products Research Society
2801 Marshall Court, Madison, Wisconsin 53705
(608) 231-1361

Furniture Design & Manufacturing
Delta Communications Inc.
400 N. Michigan Avenue, Chicago, Illinois 60611
(312) 222-2000

Furniture Today
Communications Today, Ltd.
P.O. Box 2754, High Point, North Carolina 27261
(919) 889-0113

Green Market Alert
345 Wood Creek Road
Bethlehem, Connecticut 06751
(203) 266-7209

Green Book's Hardwood Lumber Marketing Directory
Miller Publishing Co.
1235 Sycamore View
P.O. Box 34908
Memphis, TN 38184-0908

Green Book's Softwood Lumber Marketing Directory
Miller Publishing Co.
1235 Sycamore View
P.O. Box 34908
Memphis, TN 38184-0908

Hardwood Purchasing Handbook
Miller Publishing Co.
1235 Sycamore View
P.O. Box 34908
Memphis, TN 38184-0908

Home Improvement Center
Vance 400 Knightsbridge Parkway
Lincolnshire, Illinois 60069

Import/Export Wood Purchasing News
International Wood Trade Publications, Inc.
1235 Sycamore View
Memphis, Tenn. 38184
(901) 372-8280

Imported Wood Purchasing Guide
Miller Publishing Co.
1235 Sycamore View
P.O. Box 34908
Memphis, TN 38184-0908

Inside Contracting
Ceilings & Interior Systems Contractors Assn.
1800 Pickwick Avenue, Glenview, IL 60025
(312) 724-7700

Journal of Forestry
Society of American Foresters
5400 Grosvenor Lane, Bethesda, Maryland 20814
(301) 897-8720

Journal of Housing
National Association of Housing and Development Officials
Watergate Building, Suite 404, 2600 Virginia Avenue, NW
Washington, D.C. 20037
(202) 333-2020

Log Home and Alternative Housing Builder
16 First Avenue, Corry, Pennsylvania 16407-1894
(814) 664-8624

Logger and Lumberman
Dixie Publications
210 North Main, Wadley, Georgia 30477

Manufactured Home Merchandiser
RLD Group, Inc.
203 N. Wabash, Suite 800
Chicago, ILL 60601-2476
(312) 236-3528

National Hardwood Magazine
Miller Publishing Co.
1235 Sycamore View
P.O. Box 34908
Memphis, TN 38184-0908

North American Forest Products Export Directory
Miller Publishing Co.
1235 Sycamore View
P.O. Box 34908
Memphis, TN 38184-0908

New Zealand Forest Industries
Profile Publishing Limited
13 Maidstone Street, Grey Lynn
Auckland, New Zealand

Pallet Enterprise
Industrial Reporting, Inc.
10205 Pepperhill Lane, Richmond, Virginia 23233
(804) 740-1567

Plywood & Panel World
Plywood & Panel World, Inc.
610 S. McDonough St., P.O. Box 2268, Montgomery, Alabama 36197
(205) 834-4525

Potentials In Marketing
The Lakewood Building
50 South Ninth Street
Minneapolis, Minnesota 55402-3165

Remodeling Contractor
National Association of the Remodeling Industry
11 East 44th Street, New York, NY 10017
(212) 867-0121

Roofing Spec
National Roofing Contractors Association
8600 Bryn Mawr Avenue, Chicago, IL 60631
(312) 693-0700

Southern Lumberman
Greysmith Publishing, Inc.
128 Holiday Court, Suite 116, Franklin, Tennessee 37064
(615) 791-1961

State Environment Report
Business Publishers, Inc.

The Northern Logger and Timber Processor
Northeastern Loggers' Association
P.O. Box 69, Old Forge, New York 13420-0069

The Softwood Forest Products Buyer
Miller Publishing Co.
1235 Sycamore View
P.O. Box 34908
Memphis, TN 38184-0908
951 Pershing Drive
Silver Spring, MD 20910-4464
(301) 587-6300

Timber Processing
Hatton-Brown Publishers, Inc., 610 S. McDonough Street,
Montgomery, Alabama 36197
(205) 834-1170

UDM/Upholstery Design & Manufacturing
Delta Communications, Inc.
400 N. Michigan Ave.
Chicago, ILL 60611
(312) 222-2000

Wood & Wood Products
Vance Publishing Corp.
400 Knightsbridge Parkway, Lincolnshire, IL 60069

Workbench
Modern Handcraft, Inc.
4251 Pennsylvania Avenue, Kansas City, Missouri 64111

World Wood
Miller Freeman Publications, Inc.
600 Harrison Street, San Francisco, California 94107
(415) 905-2200

APPENDIX D

. . . marketing references

American Demographics
Box 68
Ithaca, NY 14851

American Statistics Index
Congressional Information Service, Washington, DC
Guide to all statistical publications of the federal government

Ayer Directory of Publications
Ayer Press, Fort Washington, PA
Lists over 22,000 newspapers and magazines.
Published annually

Basic Marketing, E. Jerome McCarthy, 1971
Homewood, Ill: Richard D. Irwin, Inc.

Better Homes and Gardens
1716 Locust Street.
Des Moines, Iowa 50336

Boardroom Reports
330 W. 42 Street.
New York, New York 10036

Business Marketing
220 E. 42nd. St.
New York, NY 10017

Business Periodicals Index
H.W. Wilson Company, New York, NY
The standard subject index to articles in business and trade periodicals. Monthly with cumulative updates. Price based on periodical subscription holdings of subscriber institution.

Classified Exchange
Miller Publishing Co.
1235 Sycamore View
P.O. Box 34908
Memphis, TN 38184-0908

Directory of Industry Data Sources—U.S. and Canada
Ballinger/Harfax, Cambridge, MA
Identifies and describes published research reports, surveys, data bases, and reference material on 60 industries.

Directory of Online Databases
Cuadra Associates, Inc., Santa Monica, CA
Brief descriptions of bibliographic, full text, and numeric computerized data bases. Access to data bases is individually priced by producers. Available in printed or online forms.

Directory of U.S. and Canadian Marketing Surveys and Services
C.H. Kline and Company, Fairfield, NJ
Brief description of over 3,000 available multi-client market research reports. Published biannually.

Entrepreneur
P.O. Box 50368
Boulder, CO 80321-0368

Entrepreneur's Marketing Guide, Roger F. Smith, 1984
Reston Publishing Company, Inc.: Prentice-Hall
Reston, Virginia 22090

Essentials of Marketing, Joel R. Evans and Barry Berman, 1984
MacMillan Publishing Company
866 Third Avenue, New York, N.Y. 10022

Export Today
P.O. Box 6072
Syracuse, New York 13217

Findex: Directory of Market Research Reports, Studies, and Surveys
FIND:SVP, New York, NY
Brief descriptions of over 7,000 marketing reports.
Published annually with supplements.

Forbes
60 5th Ave.
New York, New York 10011

Fortune
Time & Life Building
Rockerfeller Center
New York, N.Y. 10020-1393

Good Housekeeping, Hearst Corp.
959 Eighth Avenue
New York, New York 10019

Harvard Business Review
Harvard Business School
Boston, MA 02163

How to Check Out Your Competition,1987
John M. Kelly
John Wiley & Sons
New York, N.Y.

Inc. Magazine
Inc. Publishing Corp.
38 Commercial Wharf
Boston, MA 02110

Industrial Marketing Management, Robert W. Haas, 1982
Kent Publishing Company
20 Providence Street
Boston, Massachusetts 02116

International Contract Adviser
Kluwer
Law and Taxation Publishers
6 Bigelow Street
Cambridge, MA 02139 U.S.A.

John Naisbitt's Trend Letter
1101 30th St. NW
Washington, DC 20007

Ladies' Home Journal, Charter Corp.
3 Park Avenue
New York, New York 10016

Marketing, David L. Kurtz and Louis E. Boone, 1981
The Dryden Press
New York, N.Y.

Marketing Analysis and Decision Making, Burton H. Marcus and Edward M. Tauber, 1979
Little, Brown and Company
Boston, Massachusetts

Marketing Definitions: A Glossary of Marketing Terms, American Marketing Association, 1960

Marketing Handbook, Edwin E. Bobrow and Mark David Bobrow, 1985
Dow Jones-Irwin
Homewood, Illinois 60430

Marketing on a Shoestring, 1988
Jeffrey P. Davidson

John Wiley & Sons
New York, N.Y

Marketing Strategy and Plans, 1989
David J. Luck, O.C. Ferrell, George H. Lucas, Jr.
Prentice Hall
Englewood Cliffs, N.J.

MasterGuide
American Institute of Architects
PacTel Publishing and American Institute of Architects Service Corporation
1735 New York Avenue, N.W.
Washington, D.C.

MaxiMarketing—The New Direction in Advertising, Promotion, and Marketing Strategy, Stan Rapp and Thomas L. Collins,1987
McGraw Hill, Inc.,
1221 Avenue of the Americas
New York, N.Y. 10020

McCall's
230 Park Avenue
New York, New York 10017

Million Dollar Directory
Dun & Bradstreet
New York, NY
Information on companies with assets of $500,000 or over
Published annually, 3 volumes.

Moody's Manuals: Industrials; Bank & Finance; Public Utilities; Transportation; Municipal and Government
Moody's Investor Service
New York, NY
Reports on publicly held companies and publicly financed government obligations. Published annually.

Off-the-Shelf
Publications, Inc.
2171 Jericho Turnpike
Commack, New York, USA 11725

Positioning: The Battle for Your Mind, Al Ries and Jack Trout, 1981
McGraw-Hill, Inc.
1221 Avenue of the Americas
New York, N.Y. 10020

Predicasts F&S Index United States
Predicasts, Cleveland, OH
Index to trade and business periodical articles.
Also publishes F&S Index International, F&S Index Europe.
Monthly with cumulative updates.

Public Hearings Procedures and Strategies: A Guide to Influencing Public Decisions, Jean Mater, 1984
Mater Research Institute
P.O. Box "O", Corvallis, Oregon, 97339

Standard & Poor's Register of Corporations, Directors and Executives
Standard & Poor's Corporation, New York, NY
Data on over 30,000 corporations. Published annually.

Successful Marketing for Small Business, William A. Cohen and Marshall E. Reddick
AMACOM
135 W. 50th St.
New York, NY 10020

Survey of Buying Power
Sales & Marketing Management Magazine, New York, NY
Data on population, income and retail sales. Published annually in two parts.

Survey of U.S. Industrial and Commercial Buying Power
Sales & Marketing Management Magazine, New York, NY
Statistical presentation of value of shipments by industry classification (SIC)

Sweet's Catalog File
Sweet's Division
McGraw-Hill Information Systems Company
1221 Avenue of the Americas
New York, NY 10020

The Asia Letter
G.P.O. Box 10874
Hong Kong

The Encyclopedia of Associations
Gale Research, Detroit, MI
Directory of over 16,000 trade and professional organizations. Published annually with supplemental updates.

Top Management Strategy, 1980
Benjamin B. Tregoe, John W. Zimmerman
Simon and Schuster
New York, N.Y.

Thomas Register of American Manufacturers
Thomas Publishing Company, New York, NY
Lists over 100,000 manufacturing companies.
Published annually, 18 volumes.

INDEX